Heinkel He 162 "White 2" of I./JG 1 as an Allied war prize. The photo is believed to have been taken in Paris.

Heinkel He 162
"Volksjäger"
Manfred Griehl

Like almost every other German aircraft manufacturer, in 1943 Heinkel began design work on a single-engine turbojet-powered fighter. The objective was to develop a light jet fighter for the *Luftwaffe*. The question of whether the He 162 was to supplement the Me 262 or be its replacement remained unanswered for the time being. The principal advantages of a machine with greater ease of manufacture than the Me 262 were reduced consumption of materials and greater use of non-strategic materials (wood). Plans called for huge production numbers and the earliest possible start of quantity production.

Beginning in early 1944 the Heinkel design bureau in Vienna (Heinkel South) began devoting more of its efforts to this new project. By 10 July 1944 this resulted in the *Projekt P.1073*, a "fast jet fighter" powered by one or two He S11 turbojet engines, or by Jumo 004C turbojets.

Nineteen further project studies were derived from this design by October 1944. The 18th and 19th were the direct predecessors of the He 162. The goal was and remained the creation of a fighter aircraft of simple construction, an "expendable or simple fighter."

At the beginning of July 1944 the RLM issued a requirement for the development of a fast jet aircraft under the designation *"1 TL – Jäger"* to be completed within eight to ten weeks. A mockup was to be completed by 1 October 1944, the first prototype had to be ready to fly by 1 December 1944, and quantity production was to begin on 1 January 1945.

Thanks to Heinkel's preliminary work, the company was able to produce a promising study in a few days. As early as 12 July 1944 *Oberst* Knemeyer (Aircraft Development Group) expressed his general approval of the project. It appeared that Heinkel was on the right track.

On 8 September 1944 *Oberst* Diesing, the Chief of Technical Air Armaments (Chef TLR), issued the guidelines for the *"1 TL – Jäger"* under the designation *"Volksjäger"* (Peoples Fighter). The design bureau responded with a modified technical description of the P.1073 fighter, which met almost all of the criteria of the design specification.

Following a brief examination of the documents, on 13 September 1944 a contract was issued which was to be pursued "with the utmost vigor." As development of the He S11 power plant had been delayed, the design was modified to accept the initially unreliable BMW 003. Production of the Jumo 004 turbojet was reserved for the Me 262 A-1a.

On 15 September 1944, however, the Heinkel design was rejected because its range and endurance appeared to be insufficient, its fixed armament was too weak, and also because of apparent difficulties in achieving large-scale production. In spite of Dir. Francke's fierce objections, the *Chef TLR* favored Blohm & Voss' design, the P.211. The Blohm & Voss fighter continued to be favored in the subsequent conferences, however, no final decision was made. At this point the existing P.1073 mockup came into play. At a conference chaired by *General-Stabsing.* Lucht on 21 September 1944 there was discussion of the progress made by Heinkel. The balance was beginning to shift.

On 22 and 23 September 1944 an improved cockpit mockup of the P.1073, which had been assigned the designation He 500 on 25 September, was shown to the Development High Commission (EHK), which was chaired by General-Stabsing. Lucht.

Only a few minor changes were required to the Heinkel design. With several problems remaining to be solved in the development of the Bv 211, on 29 September 1944 the unexpected order was issued for the "immediate start of work on the Heinkel design."

This He 162 was tested in Mont de Marsan after the war.

The *General der Jagdflieger*, *General* Galland, and the Chief of the *Luftwaffe* General Staff, *General der Flieger* Kreipe, both had reservations about the *"Volksjäger."* Both men were of the opinion that the only way to regain even partial air superiority was to allocate all production capacity to the Me 262 A-1a. These views were not shared by *Reichsmarschall* Göring or *Hauptdienstleiter* (HDL) Karl-Otto Saur, chief of the armaments staff. In spite of all the doubts about the Heinkel design, one conference followed another at all levels.

On 30 September 1944 it was clear to Heinkel that the first production variant, the He 162 A-2, would not be available before February 1945. Despite the design's simplicity there were still numerous detail problems to iron out. Installation of the single BMW 003 turbojet, the configuration of the wings and empennage, the undercarriage layout, and the fuel tank arrangement were all changed a number of times. Charges by Professor Messerschmitt, who spoke out against the *"Volksjäger,"* had to be countered by the Heinkel company.

On 3 October the mockup of the He 500, which was by now designated the He 162 *"Volksjäger,"* was inspected again. The Chef TLR requested further changes. Heinkel was ordered to redesign the canopy jettisoning system and add fittings for take-off-assist rockets and a high-altitude breathing system. Four days later *Major* Grasser and *Fliegerstabsing.* Rauchensteiner (both members of the *General der Jagdflieger*'s staff) also inspected the He 162 mockup.

In the meantime, the first order had been placed for 1,000 machines, even though not a single prototype had yet flown.

Personnel in the design bureau often worked 90 or 100 hours per week, sleeping between the drawing boards. Wind tunnel tests began at the AVA Göttingen in September, and these were completed after much effort. Thirty-five percent of the drawings for the BMW 003-powered He 162 were completed by 20 October 1944.

In spite of the small number of available HeS 011 prototype engines, there were still demands for a version of the He 162 powered by the HeS 11 A-1. Drawings were to be ready by 1 November 1944, and design work was to be completed by 10 December. Optimists calculated that the first prototype would be ready by 20 January 1945 with a second to follow on 1 February.

At the end of October 1944 the He 162 development contract was given the *Führer*s highest priority, and *Generalkommissar* Phillipp Kessler personally took charge.

On 30 October 1944 all participants in the program were told: "The Heinkel firm must develop the aircraft which will sweep the *Terrorflieger* from our skies." But the entire infrastructure for this vast undertaking did not as yet exist. All available transport capacity was readied, work was begun on a narrow-gauge railway to Hinterbrühl near Mödling (Languste factory), and living quarters were built. 2,400 prisoners were assigned to the Languste site. By this time the SS had become involved in the project. The wings were to be built in several groups of production sites with the support of the SS. The production rate for the *"Volksjäger"* was increased from 1,000 to 2,000 machines per month. The SS had sufficient numbers of workers with which to realize these ambitious plans. The head of the *SS-Hauptamt* (Main Office) assigned *SS-Hauptsturmführer* Dr. Kurt May the task of immediately solving all questions concerning the procurement and processing of wood. Furthermore, at the beginning of November 1944 the underground Mittelwerke factory in Kronstein (near Nordhausen) was instructed to manufacture a further 1,000 He 162s and 2,000 BMW 003 power plants.

Damaged He 162 "27" on the airfield near Bernburg.

By the beginning of November 1944 the first frames for the forward fuselage of the He 162 V1 (later M1) were in the jigs. The first wing had been built hastily and therefore inaccurately. Delivery of the tail section and missing items of equipment was imminent. Construction of the He 162 V2 had begun, and final assembly started on 7 December 1944. Meanwhile, delays in the delivery of wings and tail sections delayed the construction of additional machines considerably.

On 6 December 1944 Dipl.-Ing. Gotthold Peter, Heinkel South's leading pilot, took the He 162 V1 (M1) into the air for the first time; however, the second flight, a demonstration for *Generalkommissar* Kessler and *Oberst* Diesing (*Chef TLR*), resulted in disaster. With the aircraft at 735 kph, the leading edge of the starboard wing separated and the aileron and wingtip broke away, damaging the tail surfaces. The aircraft went out of control and crashed near Fischamend, 400 meters beyond the airfield boundary. The pilot, Gotthold Peter, was killed.

An immediate investigation was begun to determine the cause of the tragic crash. The He 162 V3 was used for static vibration tests. The wood processing and gluing was checked again, and consideration was given to replacing the wooden wing with a metal one.

The next prototypes to be completed underwent intense scrutiny before they were allowed to join the flight test program. In mid-December nine faults were found with the He 162 V2, and flight testing was temporarily suspended. The *Chef-TLR* appointed an investigating commission which examined the entire He 162 design for structural strength and flight safety from 16 to 20 December 1944.

On 22 December 1944 Director Carl Francke made the first flight in the He 162 V2. Fl.Stabs-Ing. Bader subsequently flew the machine to a speed of 500 kph. He complained about the machine's inadequate rudder and aileron forces and the behavior of the power plant. Otherwise his assessment was quite satisfactory. By 15 January 1945 pilots Schuck and Kennitz had been checked out on the He 162 M2 and undertook further flights.

The He 162 was earmarked for radio trials, and Heinkel South began flight trials on 16 January 1945. By 22 January the machine completed thirteen flights with a total duration of 80 minutes.

A reworking of the design resulted in reinforcement of the tail section and wing by the end of December. The first of the improved machines was the He 162 V4. It was ferried to Heidfeld on 28 December 1944, where fifty airframe defects were found. As a result, the machine's first flight did not take place until 16 January 1945. The V4 only completed ten flights with a total duration of three hours before it was

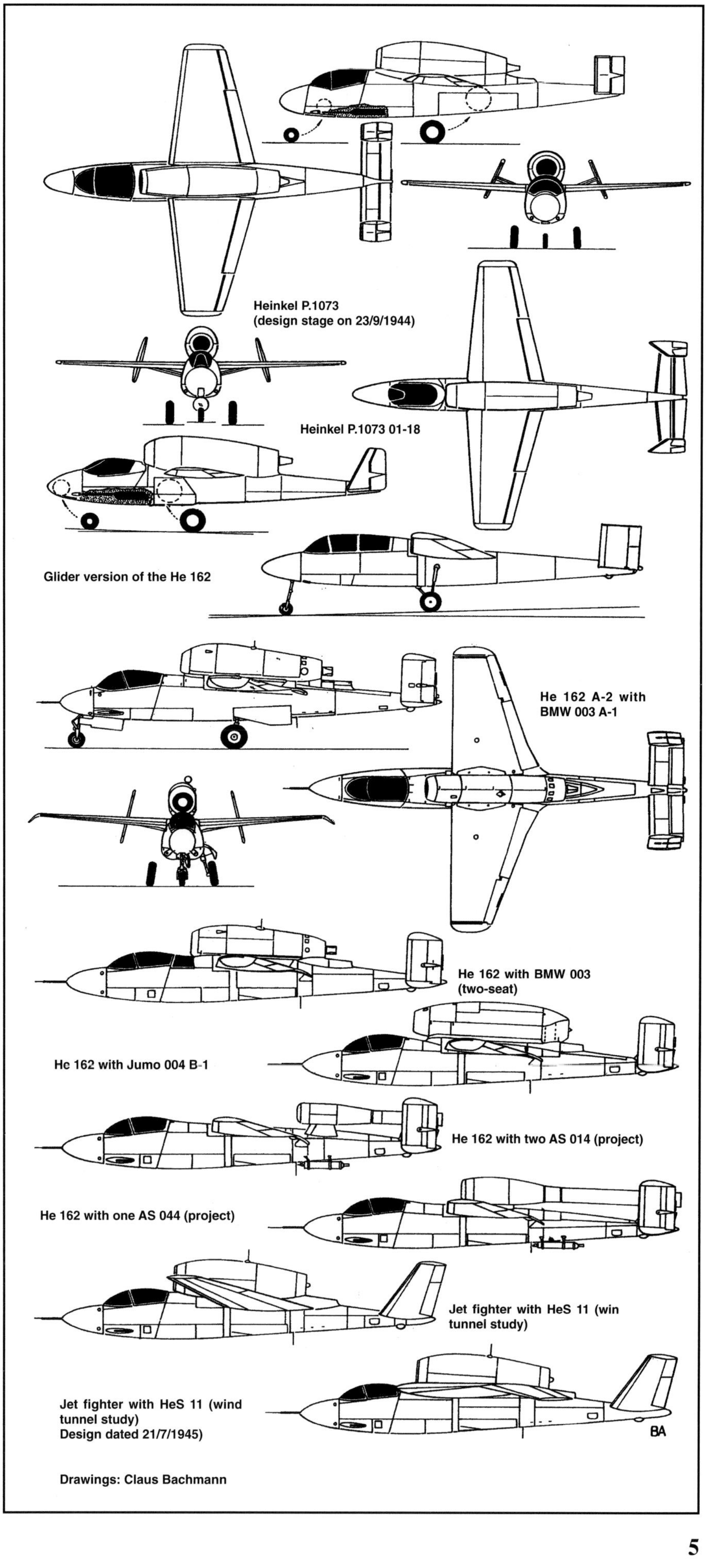

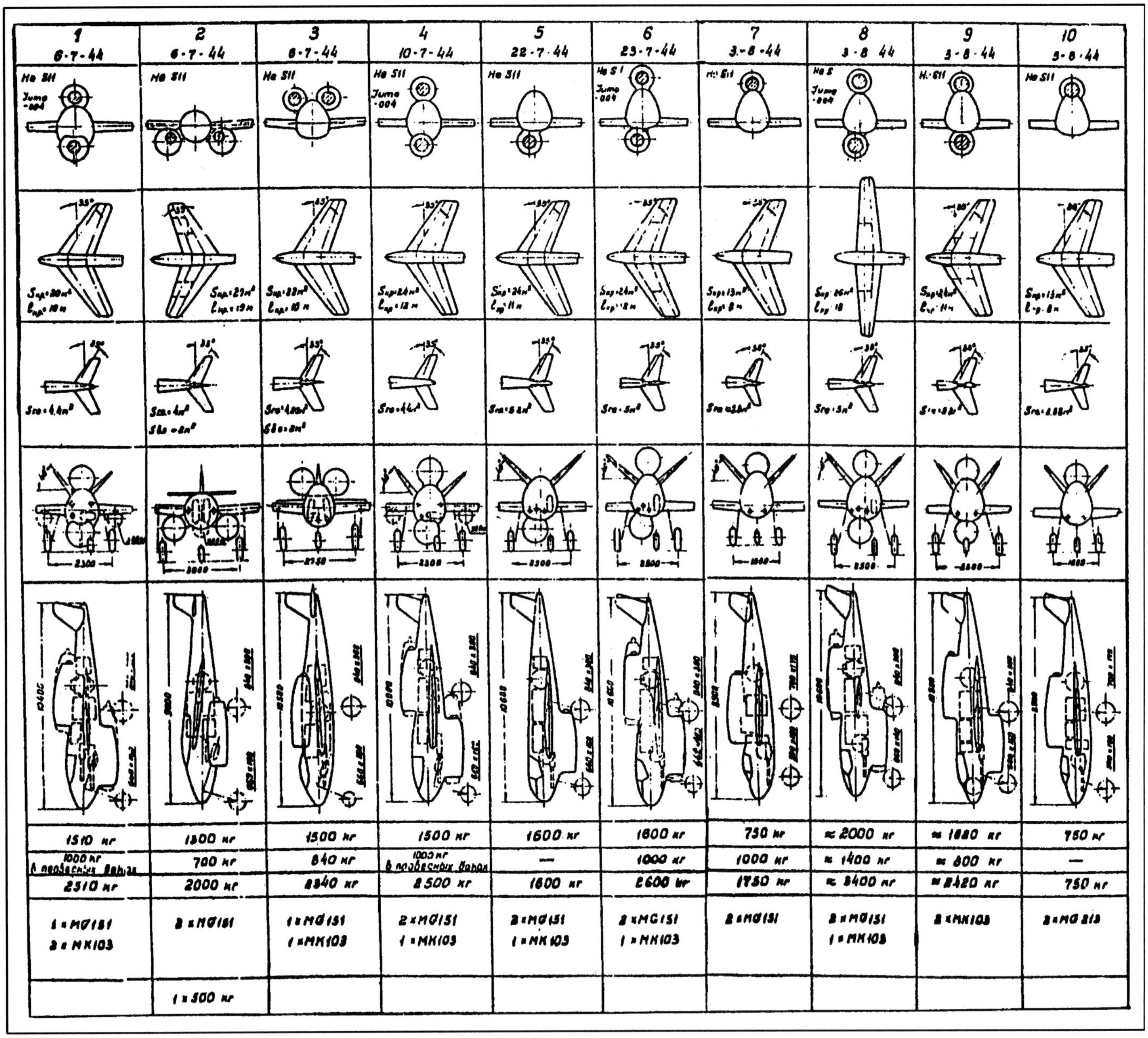

involved in a minor crash later in January 1945.

On 22 January 1945 the He 162 M6 and the first pre-production machine (He 162 A-01) were selected to train further pilots. The next day Schuck, Francke, Bader, Wedermeyer, Seling, and Bawolka flew the sixth prototype. It was determined that nosewheel damping was not necessary and could thus be deleted from production aircraft. The M6 crashed on 4 February 1945, killing Oblt. Wedermeyer.

The next machine, the He 162 M7, was fitted with a braking parachute as a safety measure for high-speed flights. Furthermore, the M3, M4, and the first A-0 were to be used to investigate stability about the normal axis, the rolling moment, and various tail section configurations in flight.

By 30 January 1945 the M2 to M7 prototypes were in the test program, as well as the first three "Zero-Series" (pre-production) machines. At the beginning of February 1945 ground fog halted the flying program briefly, providing time to repair the M6, which was damaged after a nosewheel failure, and to replace the ailerons on the A-02. Following installation of the braking parachute the M7 was cleared for flight trials.

By 5 February 1945 the prototypes had completed 63 flights totaling ten hours and fifty-seven minutes. At that time a total of fifteen pilots had flown the He 162, including several with only a B2 rating, meaning they had little experience on front-line machines.

Soon afterwards *Flugzeugbau-meister* Full reached 800 kph (Mach .65) for the first time at an altitude of 8,000 m. Serious vibration developed, after which the engine quit. Full was obliged to make a forced-landing in snow-covered terrain, sustaining minor injuries. In an effort to improve stability, the He 162's fuselage was lengthened. Furthermore, downward-canted wingtips were introduced to reduce the type's tendency to roll. It was hoped that enlarged horizontal tail surfaces and modified fins and rudders would finally result in an acceptable service aircraft.

With the He 162's most serious weak points identified, on 14 February 1945 work could begin to modify all of the prototypes and pre-production aircraft built to date.

In mid-February 1945, *Flugzeugbaumeister* Full reached a speed of 880 kph while flying the M3, the first He 162 to be so modified. The prototype proved completely stable in all flight attitudes.

Beginning at the end of February 1945, wind incidence was increased and a fairing was added to the wing-fuselage join. At this time the M3 crashed as a result of an engine fire while being flown by *Flugzeugbaumeister* Full. He ejected successfully,

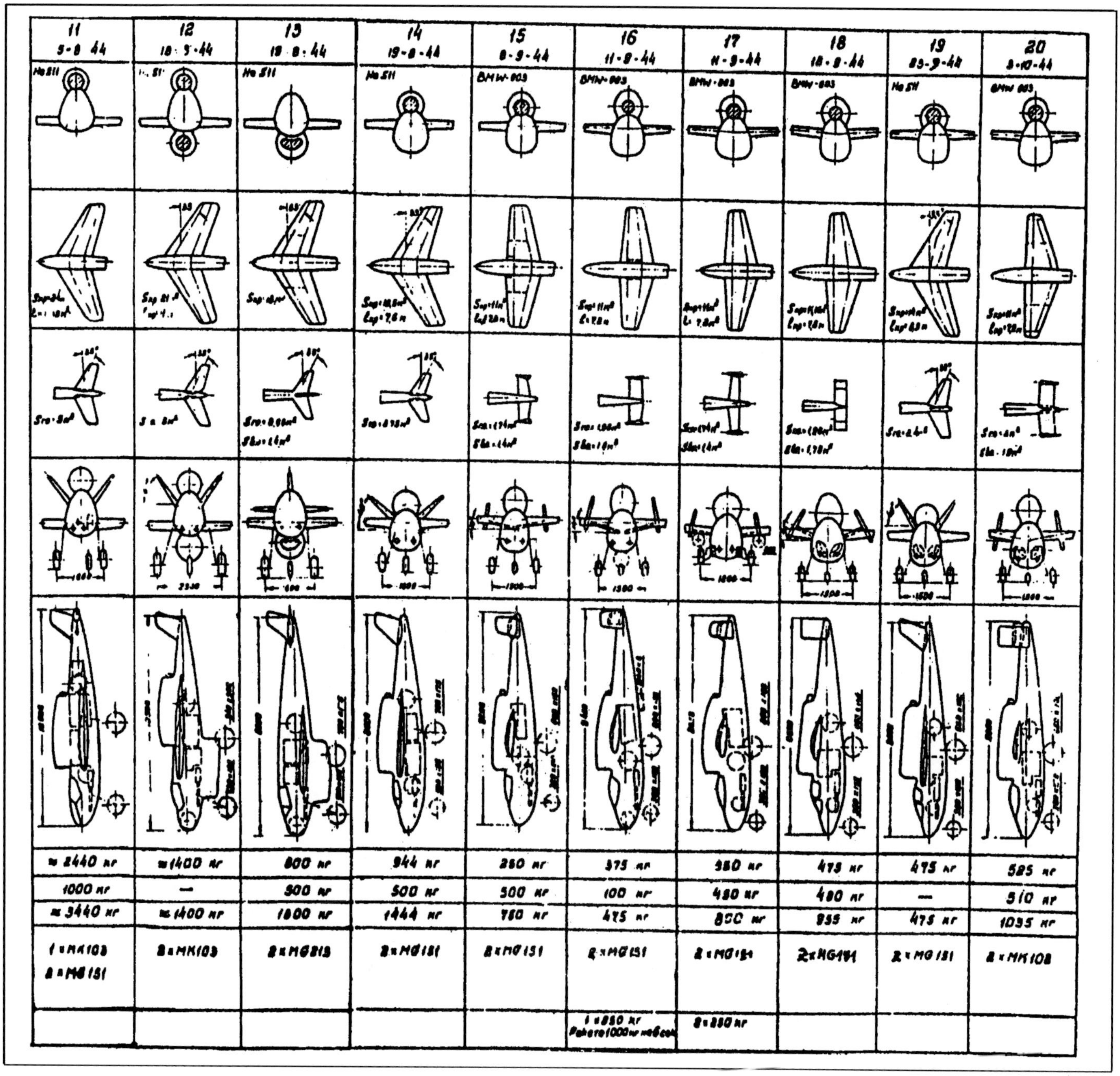

however, at a height of 200 meters there was insufficient time for his parachute to deploy.

By 25 February 1945 Heinkel South had completed 166 flights with a total flight time of 40.5 hours. Development of the *"Volksjäger"* slowed perceptibly as a result of the rapidly deteriorating supply situation and the frequent air raid alerts. On 2 March 1945 the M25, a machine with a lengthened fuselage, sustained 60% damage in a crash. Soon afterwards a second He 162 (pilot Denzin) sustained minor damage.

By 11 March 1945 the number of He 162 flights climbed to 211 (51 hr. 13 min.). One day later the M8 prototype was involved in a crash when pilot Fw. Wanke touched down too early while landing. The aircraft overturned and caught fire. Wanke survived with serious burns and bruises. On 14 March 1945 Uffz. Daus (or Tautz) of 2.JG 1 (Heidfeld Interceptor *Staffel*) was killed when his aircraft collided with several containers lying beside the runway.

Heinkel South's last weekly report is dated 26 March 1945, and it records that

He 162 with wings, nose, and engine removed.

He 162 Prototypes

Prototype	Scheduled First Flight	Actual First Flight	Werk-Nr./Code	a) Test Pilot b) Purpose c) Fate
He 162 M1	02/12/1944	06/12/1944	200001 VI + IA	a) Flugkapitän Dipl.-Ing. Gotthold Peter b) Handling trials with Heinkel South c) Crashed on 10/12/1944 (100%), Flugkapitän G. Peter killed.
He 162 M2	17/12/1944	22/12/1944	200002 VI + IB	a) Dir. Carl Francke b) Handling, stability trials and performance assessment, static weapons trials with MK 108 (continuous fire), later static tests c) Last reported 23 February 1945
He 162 M3	20/12/1944	16/01/1945	200003 VI + IC	a) Kemnitz b) Stability and radio trials (radio compass), testing of new larger tail surfaces and undercarriage with shortened retraction struts, installation of extended wing, finally stability trials. c) Crashed on 25 February 1945 (100%), Flugzeugbaumeister Full killed.
He 162 M4	04/01/1945	16/01/1945	200004 VI + ID	a) Schuck b) Stability and roll tests, MG 151 ammunition, airframe strength tests, strengthened wings and ailerons plus new ailerons, gunsight installed.
He 162 M5	27/12/1944	- - -	200005	a) Not flown b) Used as static test airframe from December 1944 c) Last reported 15 February 1945
He 162 M6	17/01/1945	23/01/1945	200006 VI + IF	a) Schuck b) Longitudinal stability measurements, control surface effectiveness, weapons trials with two MK 108 (A-1 version) c) Crashed on 4 February 1945 (100%), Oblt. Wedemeyer (factory test flight)
He 162 M7	- - -	22/01/1945	200007 VI + IG	a) Schuck b) Vibration and braking chute trials (A-2 version) c) Damaged in a Heinkel South hangar on 2 March 1945
He 162 M8	24/01/1945	27/01/1945	200008 VI + IH	a) not known b) Trials with larger nosewheel plus weapons trials with 2 x MG 151/20 (A-2 version), installation of new lighter undercarriage c) Landing accident on 12 March 1945 (100%), Fw. Wanke (injured)
He 162 M9	15/02/1945	- - -	200009 VI + II	a) not flown b) Planned as two-seat trainer with BMW 003, not completed. c) Airframe probably blown up near Vienna on 1 April 1945.
He 162 M10	28/02/1945	- - -	200010	a) not flown b) Planned as two-seat trainer with BMW 003, not completed. c) Airframe probably blown up near Vienna on 1 April 1945.
He 162 M11	15/01/1945	- - -	220017 WA + ??	a) not flown b) Prototype with Jumo 004 and MK 108 weapons system, converted in Mödling on 14 February 1945. c) Destroyed in air raid on Heinkel South in March 1945.
He 162 M12	01/02/1945	- - -	220018 WA + ??	a) not flown b) Prototype with Jumo 004 and MG 151/20 Weapons system, converted in Mödling on 14 February 1945. c) Destroyed in air raid on Heinkel South in March 1945.
He 162 M13	not assigned			
He 162 M14	- - -	15/02/1945	- - -	a) - - - b) First prototype with HeS 11 turbojet engine, construction set back because of non-availability of engine, MK 108 armament planned. c) Not built
He 162 M15	- - -	28/02/1945	- - -	a) - - - b) Second prototype with HeS 11, construction set back because of non-availability of engine, MG 151 armament planned. c) Not built.
He 162 M16	- - -	- - -	- - -	a) - - - b) Third airframe for training aircraft (survey dated 18/02/1945). c) Not built.
He 162 M17	- - -	- - -	- - -	a) - - - b) Fourth airframe for training aircraft (survey dated 18/02/1945). c) Not built.
He 162 M18	10/01/1945	24/01/1945 A-01	220001 VI + IK	a) Not known b) Roll tests, takeoff and landing characteristics, radio tests, prototype of the He 162 (BMW 003 E-1), endurance tests, recording camera installed. c) Last reported on 14 March 1945.
He 162 M19	16/01/1945	28/01/1945	220002 VI + IL	a) Not known b) Handling trials, new wing incidence (1 degree), additional rudder tests and rudder deflection measurements, 20-degree vee-tail with ultimate mass balance. c) Landing accident on 14 March 1945, Uffz. Daus (2./JG 1) killed.
He 162 M20	18/01/1945	10/02/1945	220003 VI + IM A-03	a) Not known. b) Handling trials with improved control system and simplified undercarriage, tail section of He 162 M19, crashed on 24 February 1945 (Dir. Francke). c) Munich-Riem airport on 1 May 1945.
He 162 M21	22/02/1945	22/02/1945	220004 VI + IN A-04	a) Gerhard Gleuwitz b) Performance tests to 955 kph, also weapons trials with MG 151/20.

				c) Minor crash on 7 March 1945 (20%), pilot not injured.
He 162 M22	24/01/1945	25/02/1945	220005 VI + IO A-05	a) Not known. b) First prototype with improved wing-fuselage join, longitudinal stability measurements. c) Minor crash on 14 March 1945 (15%), Fw. Denzin.
He 162 M23	27/02/1945	19/03/1945	220006 VI + IP A-06	a) Gerhard Gleuwitz b) Second prototype with improved wing-fuselage join and wing incidence of the M19 (2 degrees). c) At Munich-Riem airfield on 1 May 1945
He 162 M24	27/01/1945	20/03/1945	220007 VI + IQ A-07	a) Gerhard Gleuwitz b) General flight trials and roll tests with Heinkel South. c) Destroyed by enemy action in March 1945.
He 162 M25	29/01/1945	17/02/1945	220008 VI + IR A-08	a) Gerhard Gleuwitz b) First prototype with fuselage lengthened by .45 m and MK 108 armament, CG measurements, last used to measure climb performance. c) Landing accident on 2 March 1945 (60%), pilot Pawolka unhurt.
He 162 M26	30/01/1945	17/02/1945	220009 VI + IS A-09	a) Not known b) Second prototype with fuselage lengthened by .45 m and MK 108 armament. c) Landing accident on 12 March 1945, pilot Gleuwitz.
He 162 M27	01/02/1945	01/02/1945	220010 VI + IT A-010	a) Not known b) Third prototype with fuselage lengthened by .45 m and MK 108 armament, damaged 2 March 1945 c) On 12 March 1945 in Heidfeld for airframe strengthening.
He 162 M28	01/02/1945	18/02/1945	220011 VI + IU A-011	a) Not known b) Prototype with lengthened fuselage and MK 108 armament (reserve aircraft). c) Fate not known.
He 162 M29	08/02/1945	19/02/1945	220012 VI + IV A-012	a) Not known. b) Prototype with lengthened fuselage and MK 108 armament (reserve aircraft). c) Fate not known.
He 162 M30	08/02/1945	24/02/1945	220013 VI + IW A-013	a) Not known. b) Reserve aircraft for trials with EZ 42 sight. c) Fate not known (probably to JG 1).
He 162 M31	08/02/1945	24/02/1945	220014 VI + IX A-014	a) Not known. b) Weapons trials, installation of EZ 42 c) Fate not known (probably to JG 1)
He 162 M32	Feb. 1945	15/03/1945	220020 VI + IY A-020	a) Gerhard Gleuwitz b) Reserve aircraft for weapons trials. c) Destroyed in air raid 14 Feb. 1945.
He 162 M33	Feb. 1945	14/03/1945	220021 VI + IT A-2	a) Gerhard Gleuwitz b) Reserve aircraft for weapons trials. c) Fate not known (probably to JG 1).
He 162 M34	Feb. 1945	22/03/1945	220022 - - - A-2	a) Gerhard Gleuwitz b) Reserve aircraft for weapons trials c) Fate not known (probably to JG 1).
He 162 M35	21/02/1945	23/03/1945	220023 - - - A-2	a) Gerhard Gleuwitz b) Third prototype with modified wing incidence (1 deg.), converted in Languste factory. c) Fate not known (probably to JG 1).
He 162 M36	21/02/1945	- - -	220024 - - - A-2	a) Not known. b) Fourth prototype with modified wing from Languste factory. c) Fate not known (probably to JG 1).
He 162 M37	21/02/1945	- - -	220025 - - - A-2	a) Not known b) Fifth prototype with modified wing incidence (2 deg.), converted in Languste Factory. c) Fate not known (probably to JG 1).
He 162 M38	Feb. 1945	- - -	220026 - - - A-2	a) Not flown. b) Reserve test aircraft. c) Destroyed in air raid on Heidfeld on 14 February 1945.
He 162 M39	March 1945	- - -	220027 - - - A-2	a) Not known. b) Reserve test aircraft. c) Slightly damaged in air raid on 14 February 1945.
He 162 M40	March 1945	- - -	220028 - - - A-2	a) Not known. b) Reserve test aircraft. c) Slightly damaged in air raid on 14 February 1945.
He 162 M41	March 1945	- - -	220029 - - - A-2	a) Not known. b) Reserve test aircraft. c) Not completed.
He 162 M42	March 1945	- - -	220030 - - - A-2	a) Not known. b) Reserve test aircraft. c) Not completed.

259 flights were completed on the *"Volksjäger"* at Heidfeld. The machines logged more than 65 hours in the air. After numerous modifications, the last affecting the fuel system, the *"Volksjäger"* was almost ready for operations.

By then Soviet units were approaching Vienna. On 30 March 1945 Director Francke went to Hauptdienstleiter Saur's office seeking a transfer order to Bad Gandersheim. This arrived twenty-four hours later, and on 1 April 1945 a special train departed for the Harz Mountains. The train was then halted for three full days in Eger before being rerouted to Jenbach in Tyrol. It arrived there during the night of 5-6 April.

A few hours earlier the Red Army had occupied the Languste factory in the Mödling area, and one day later the area around Schwechat, which included the company airfield at Heidfeld. Several days

He 162 A-2 of I./JG 1 at the end of the war (May 1945).

earlier the factory directors had begun moving some of the jigs and tools for the He 162 in the Heinkel factory to Jenbach. The parts subsequently arrived in Lent near Salzburg.

The design bureau was moved from Jenbach to Landsberg/Lech, where it arrived on 14 April 1945. barely three weeks later, on 14 April 1945, American troops reached the Lech. Development of the *"Volksjäger"* ended for good.

In the Vienna area, meanwhile, Soviet units captured numerous documents and sufficient undamaged components to assemble at least three He 162s for test purposes. Beginning in 1946 the captured aircraft were tested at the research institute of the NII VVS, however the type was not adopted by the Soviets.

Production of the He 162 A-2 was supposed to be carried out as a so-called "Forced Measure." As a result, the few remaining resources could be assigned to the production program without regard to other areas. This applied to the necessary construction materials, transport, and labor forces.

After the ultimate decision to mass produce the *"Volksjäger"* was made on 29 September 1944, the Reich Ministry for Armaments and War Production (RMfRuK) decreed (Diary No. 999718/44):

"Authorization is granted for the immediate start-up of production of the small single-jet fighter developed by Heinkel, which the Development High Commission recommended for production after an exhaustive examination. The interim monthly production figure is set at 1,000 examples.

Furthermore, it is acknowledged that there are significant risks associated with the method of deciding to proceed from the design stage directly to quantity production."

On 12 October 1944 Generalkommissar Phillipp Kessler was placed in charge of the entire "*Volksjäger* Program." Seven days later the "Tortoise Program" was presented to the Reich Ministry for Armaments and War Production. The proposal called for 1,000 He 162s and 2,000 BMW turbojets to be built in the underground Mittelwerke factory, which had an area of 25,500 m^2. Subsequent plans were equally imposing:

Period	He 162	BMW 003
January 1945	1	10
February 1945	5	100
March 1945	15	400
April 1945	25	600
May 1945	45	1,000
June 1945	300	1,000
July 1945	600	1,000
August 1945	1,000	1,000
to April 1946	1,000	1,000

Furthermore, on 30 October 1945 it was decided that final assembly of the He 162 would also take place at the Heinkel facilities in Marienehe and Vienna and the Junkers plant in Bernburg. At the same time, subcontractors for the necessary wooden components were concentrated in three groups in the Erfurt, Stuttgart, and Vienna areas. Heinkel was to produce the metal components, with facilities in Barth/Pomerania, Pütnitz/Mecklenburg, and Stassfurt/Saxony providing one-third. The remaining two-thirds was to be produced by Junkers in its Aschersleben, Bernburg, Halberstadt, Leopoldshall, and Schönbeck plants. Furthermore, Deutsche Lufthansa was to become involved in component manufacture and later final assembly.

On 28 November 1944 a new production plan, more in keeping with the realities of the year 1945, was presented to the Reich Ministry for Armaments and War Production:

Underground production site in Egeln near Staßfurt, where the first He 162 was completed at the end of April 1945.

Period	Heinkel/ Junkers	Mittelwerke (Kunze)
January 1945	50	0
February 1945	100	0
March 1945	200	2
April 1945	300	5
May 1945	400	15
June 1945	600	25
July 1945	700	45
August 1945	1,000	300
to April 1946	1,000	500

After being shown the forecast production figures at the end of November 1944, at which time the first prototype had not yet flown, Hitler was already looking forward to the mass employment of the He 162.

After the crash of the He 162 M1 on 10 December 1944 the wings had to be strengthened, modifications made to the tail assembly and undercarriage, improvements made to the wing-fuselage join, and, later, the fuselage extended. Numerous other detail changes are not mentioned here.

In order to speed development and the start of production, Heinkel introduced two twelve-hour shifts. The first 15 sets of modified wings were completed by 21 January 1945. The absence of fins and rudders delayed completion of He 162s at Heidfeld (Schwechat near Vienna). Just twelve machines were delivered by 6 February 1945.

As a result of delays, in mid-1945 the entire He 162 schedule had to be rethought. The following deadlines were established:
a) Start of service trials: beginning of March 1945
b) Start of E-Stelle (test station) trials: beginning of April 1945
c) Completion of first service aircraft: mid-April 1945

The underground Languste factory near Mödking near Vienna.

Assembly of the first production fuselage by Heinkel-South.

Entrance to the Languste final assembly operation.

The first He 162 built at Marienehe was ready to fly on 25 March 1945. Photo: Dr. Alfred Hiller.

d) Entry into service of first He 162 Jagdgruppe: mid-May 1945

In order to ensure that these target dates were achieved, final assembly of ten He 162s was to be completed at Heidfeld (by 2 March 1945), Marienehe, Theresienfeld near Rostock (by 5 March 1945), and in Bernburg (by 10 March 1945).

As of 18 February 1945, however, all *"Volksjäger"* produced so far had to be modified as per the recently issued modification directives in order to improve flight safety. On 6 March 1945 orders were issued for additional modifications to those machines already built, which meant further delays. Consequently, deliveries of the first production He 162 As were slower to begin than expected. The first production aircraft were test flown later than planned: in Bernburg (Junkers) on 23 March 1945, Oranienburg (Heinkel) on 24 March 1945, and Marienehe (Heinkel) on 25 March 1945.

In Oranienburg the first *"Volksjäger"* was ready for its maiden flight on 24 March 1945.

Until then the general situation only allowed the delivery of improved machines from the Vienna area. On 27 March 1945 the following production aircraft existed in various stages of construction:

Bernburg (Junkers) 18 airframes without wings and engines,

Oranienburg (Heinkel) 10 airframes without wings and engines,

Marienehe (Heinkel) 5 airframes under assembly,

Ludwigslust (Heinkel) 10 airframes under assembly,

Mödling (Heinkel) 15 airframes under assembly,

Nordhausen (Kunze) 5 airframes under construction.

Of the machines listed above, approximately 20% were probably completed by mid-April 1945. Soon afterwards, on 7 April, Hitler ordered production of the He 162 and Ar 234 halted in favor of the Me 262. In the meantime, at the beginning of April Soviet troops had occupied the Vienna area (and thus Heidfeld and Hinterbrühl). The Mittelwerk was lost on 10 April 1945. From then on contact with the production sites at Rostock and Junkers was repeatedly interrupted. At Theresienfeld, however, final assembly did not collapse until the end of

A *"Volksjäger"* and its starter cart in tow behind a Hanomag prime mover.

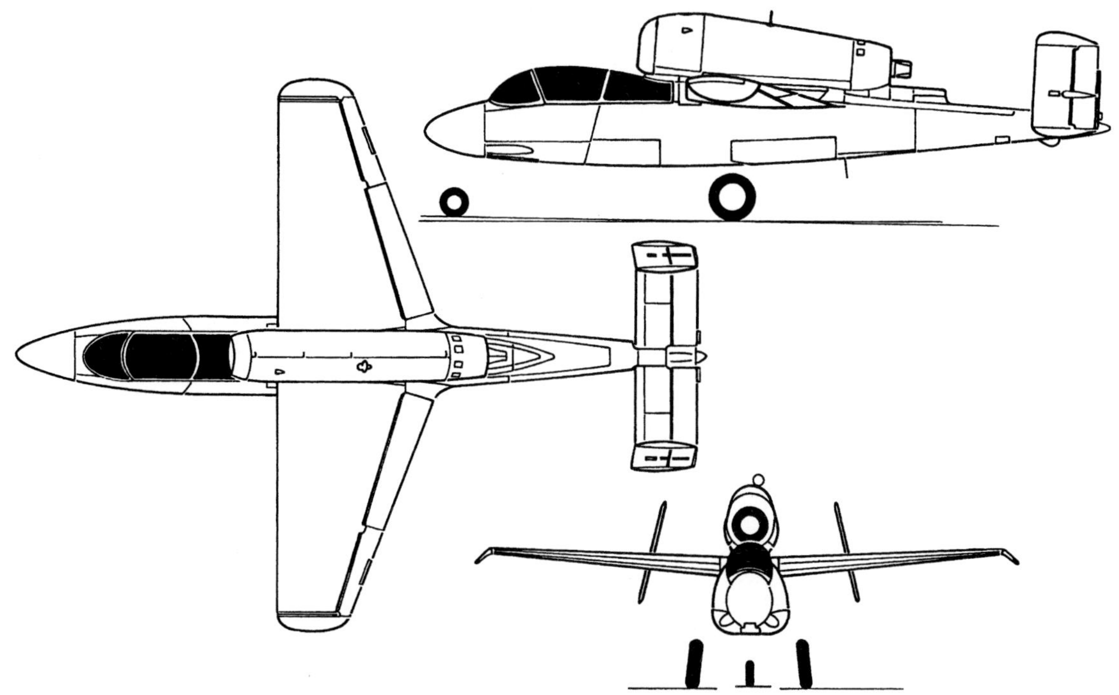

Dreiseitenansicht des geplanten doppelsitzigen Schulflugzeugs Heinkel He 162S

Three-view drawing of the planned He 162 S two-seat training aircraft.

April. In the case of Junkers this happened somewhat earlier as a result of the Allied advance.

By April 1945 "*Volksjäger*" production brought out more than 125 machines, which were built in the following production facilities:

Vienna-Schwechat (Heidfeld, Heinkel)
10 aircraft Werk-Nr. 200001 and subsequent aircraft
Hinterbrühl (Languste factory, Heinkel)
at least 20 aircraft Werk-Nr. 220001 and subsequent aircraft
Rostock-Marienehe (Heinkel))
at least 55 aircraft Werk-Nr. 120001 and subsequent aircraft
Rostock-Theresienfeld (Heinkel)
at least 20 aircraft Werk-Nr. 120001 and subsequent aircraft
Bernburg (Junkers)
at least 15 aircraft Werk-Nr. 300001 and subsequent aircraft
Nordhausen (Mittelwerk)
at least 5 aircraft Werk-Nr. 310001 and subsequent aircraft
Oranienburg (Heinkel)
1 to 5 aircraft Werk-Nr. not known

In addition to those He 162 A-2s which were completed, components for another hundred machines had been produced or were in an advanced state of construction in April 1945. During interrogation on 7 July 1945 Prof. Ernst Heinkel stated that in April 1945 alone there were 120 to 140 engineless He 162s at the Heinkel facilities in the Rostock area and Junkers in Bernburg. Some of these aircraft must subsequently have been fitted with the BMW 003, otherwise the total number of completed He 162 As would have been just 60 to 80 machines. Of the completed aircraft only a relatively small number reached the operational units, as the airframes had to be modified as per the numerous modification directives before delivery.

No aircraft were delivered to front-line units by 31 January 1945. The aircraft distribution survey by the GenQu. 6.Abt. dated 2 March 1945 was the first to include the He 162, showing 81 machines produced in February 1945. Of these 63 were under conversion, which left just 18 for the flying units. In March, the necessary conversion work meant that just 22 aircraft were delivered to the air force. Not until April 1945 did the number of "Volks*jäger*" delivered exceed the 20 mark. These machines were the ultimate version, which required no conversion work prior to delivery.

The aircraft distribution reports for the period from February to April 1945 list the following deliveries:

14/02/1945 1 He 162 Parchim Aviation Technical College
14/02/1945 1 He 162 Bayreuth Aviation Technical College
27/02/1945 16 He 162 Technical Air Armaments (TLR)
14/03/1945 1 He 162 Bayreuth Aviation Technical College
26/03/1945 2 He 162 KdE – Erprobungsstelle Rechlin
27/03/1945 1 He 162

The training glider based on the He 162.

Flugzeugüberführungsgeschwader 1
30/03/1945 1 He 162 KdE
30/03/1945 6 He 162 III./JG 1
31/03/1945 7 He 162 III./JG 1
31/03/1945 2 He 162 Stab JG 1
31/03/1945 2 He 162 Flugzeugüberführungsgeschwader 1
03/04/1945 2 He 162 Flugzeugüberführungsgeschwader 1
04/04/1945 3 He 162 Technical Air Armaments (TLR)
05/04/1945 3 He 162 Stab JG 1
06/04/1945 2 He 162 Faßberg Aviation Technical College
06/04/1945 2 He 162 I./JG 1
10/04/1945 4 He 162 Stab JG 1
11/04/1945 3 He 162 I./JG 1

Because of the military situation no further reports survived. The pilots of JG 1 collected their machines directly from the manufacturer. According to the official statistics of the GenQu. 6.Abt. 116 He 162s were delivered to the Luftwaffe. Of these, at the end of the war 31 aircraft were at Leck, an interesting find for the Allied specialists.

The latter were considerably more interested in projected developments of the He 162. The problems with the BMW 003 turbojets and the non-availability of the HeS 11 turbojet originally planned for the single-engined jet fighter forced the development department to investigate alternative solutions.

All efforts to install Jumo 004 turbojets in two He 162s were foiled by an air raid. The engines had arrived in Hinterbrühl on 14 February 1945 and were installed in the 11th and 12th prototypes. A proposal to introduce a larger wing was rejected by HDL Saur at the end of January as the conversion was too costly. This also applied to various swept wings and the vee tail, which was in the advanced design stage in 1945. The various special armaments (R4M, R50BS, *Panzerblitz*) likewise went unrealized, as did proposed gun armament packages (4 x MG 151/20, MK 213, *Rohrblock*). Proposals to fit the machine with bomb dispensers (AB 250) and SD 70 bombs remained just that.

View into the abandoned cavern of the Languste factory.

The first He 162 built by Junkers prior to its maiden flight on 24 March 1945.

Development and production start-up of the He 162 S training aircraft had just begun when the war ended.

First mention of a "glider trainer with winch tow" and a "dual-control aircraft" based on the He 162 A-2 was made in a memo dated 14 October 1944. According to Dipl.-Ing. Günther, two prototypes were flown in mid-March 1945. This is contradicted by a statement by the Chef TLR, according to which the first He 162 S made its first flight at Trebbin on 1 March 1945.

Hauptdienstleiter Saur and *Reichsjugendführer* Axtmann finally rejected the planned mass employment of the *"Volksjäger"* in the hands of the Hitler Youth. *Generaloberst* Keller, on the other hand, remained an enthusiastic supporter of the plan. A large number of He 162 S trainers was to be built by calling upon all available resources of the *Nationalsozialistische Flieger-korps* (National-Socialist Air Corps) and employing local wood-working operations. Heinkel began designing the initial training aircraft at the end of October 1944.

Cockpit of a He 162.

Heinkel He 162 with BMW 003 E-2 engine, found by Allied troops in Leck. The aircraft was later shipped to the USA.

Initial plans called for the construction of ten prototypes. As well as the single-seat glider version with the standard wing, in December 1944 work was proceeding on a version with a larger wing and two-seat cockpit. The fourth version of the training machine, a two-seater powered by a BMW 003 E-1 or E-2 turbojet, was under construction by DLH in Oranienburg at the end of 1944. Quantity production was to be undertaken by the Kleinemeyer firm.

The first young pilots from the ranks of the Hitler Youth and the National-Socialist Air Corps were supposed to be trained in "Special Courses" at Trebbin. For this purpose, in roughly mid-March 1945 three He 162 fuselages were delivered to the SAB firm in Bendorf for conversion into flight simulators.

On 1 March 1945 the first He 162 glider was test-flown by the Reich Gliding School in Trebbin. The remaining He 162 S gliders stored there as components were assembled by the end of the month. According to Georg Cordts, who took part in the course, pilots carried out seven winch launches on 28 March 1945. Twenty-four hours later the He 162 S was also flown successfully with a tow aircraft. A few days later, on 8 April 1945, Hanna Reitsch arrived at Trebbin in a Bü 181 to fly the He 162 S.

On 24 April 1945 the war put an end to the "Replacement Fighter Pilot for Special Purposes" course. The personnel were not sent to JG 1, however; instead, they were employed as ground troops by the *Division Jahn* in the defense of Berlin, seeing action in the Potsdam-Werder area.

Formation of the 1.(*Volkssturm-staffel*)/JG 1 did not take place, as the Sagan airfield had already been occupied and there were insufficient aircraft.

Although Germany was on the verge of collapse, a great deal of emphasis was placed on producing pilot notes and an illustrated manual for operation of the *"Volksjäger."* These were completed, however, none reached JG 1.

In spite of serious difficulties, the He 162 entered quantity production in December 1944 as a "Forced Measure." The production program included several Heinkel factories, Junkers, and the Mittelwerke, plus numerous small- and medium-size companies. Possibilities were limited by the military situation and the encroachment of the enemy, however, Germany's leaders grasped at straws, and one of them was the *"Volksjäger."* In December 1944 it was apparent that the first machines would be available for training and tactical trials at the beginning of 1945. On 27 December 1944 the Commander of Test Stations (KdE), in conjunction with the Chief of Technical Air Armaments (Chef TLR), proposed setting up an He 162 test detachment in Lärz. The unit was to be of *Staffel* size initially and begin work in February 1945. On 1 January 1945 the *General der Jagdflieger* (GdJ) asked the Quartermaster-General (GenQu) to expand the test detachment to *Gruppe* size. The unit would subsequently serve as a pure replacement training *Gruppe* for He 162 pilots.

One day later, however, the *Luftwaffe* High Command (OKL) decided that the test *Staffel* should be stationed near the manufacturer. At the same time, on 9 January 1945 the *I. Gruppe* of JG 200 was formed in *Luftflotte Reich*'s area of command and attached to the *General der Jagdflieger* for training. The purely technical testing was to be done by the *Stabsstaffel* of JG 200 stationed at Lärz. A short time later the formation of JG 200 was surprisingly halted.

On 10 January 1945 the Quartermaster-General authorized the formation of an

He 162s at Leck prior to the surrender. Photo: Archiv Peter Petrick.

He 162s at Leck air base in May 1945. Photo: Archiv Peter Petrick.

He 162s of I./JG 1 lined up in preparation for surrender to the Allies. Photo: Archiv Peter Petrick.

Erprobungskommando 162 (Test Detachment 162) for a duration of six months. A few days later, on 14 January 1945, 27 technical personnel were sent to the Heinkel factory in Marienehe.

Then, on 25 January 1945, a formation order was issued for the *Stab* (headquarters) of the new JG 80 with *Stabsstaffel*, headquarters company, and one *Gruppe* of three *Staffeln*, each with twelve He 162s. The *Gruppenstab* was to be equipped with four more He 162s. Two days later 1./JG 80 was ordered to move to Vienna-Aspern (He 162 delivery point) beginning on 15 February 1945. The *Stab* plus 2. and 3./JG 80 was to move to Parchim, and the *Stabsstaffel* JG 80 to Rechlin. The element of JG 80 stationed at Rechlin was to receive its aircraft directly from factories in Rostock (Heinkel) and Bernburg (Junkers).

The order for the formation for JG 80 was rescinded on 7 February 1945. Instead, an order was issued for *I. Gruppe* of JG 1 (*Stabsstaffel* and three flying *Staffeln*) to become the first *Luftwaffe* unit to be equipped with the *"Volksjäger."* In addition to training, the *Stabsstaffel* was to be responsible for tactical testing of the He 162 and was therefore subordinated directly to the *Kommando der Erprobungsstellen* (KdE).

On 11 February 1945—by order of the *Luftwaffe* Organization Staff—the *Stabsstaffel* JG 1 was assigned to the KdE. The planning called for the *I. Gruppe* of JG 1 to be brought up to strength with personnel from II. and III./JG 1. At the same time, a telex from *Luftwaffenkommando 6* reached I./JG 1 with orders for the unit to be withdrawn from operations immediately to rest and reequip. The unit's Fw 190s were to be given to II./JG 1. *II. Gruppe* was also supposed to send an advance detachment

An operational machine of JG 1 ready for transport. Photo: Archiv Manfred Griehl.

Major Zober's He 162 A-2 (Stab I./JG 1) parked at Kassel. Photo: R.P. von Lutz.

to Vienna-Aspern to collect He 162s, however, the end of the war prevented this.

The entire *Geschwader* area had to be prepared for the *"Volksjäger"* as quickly as possible. On 25 January 1945 *Jagdgeschwader 1*'s situation was not especially good. *I. Gruppe*, which had been transferred back under the command of Oblt. Demuth, had just 13 Fw 190 A-8s and A-9s. The remnants of *II. Gruppe* followed a short time later, however III./JG 1 was almost completely overrun by Soviet forces in East Prussia. The remaining Fw 190s were used as fighter-bombers until the beginning of 1945, when II./JG 1 was transferred to Rostock-Marienehe, from where it continued its operations against the Red Army. Meanwhile, on 9 February 1945 I./JG 1's train arrived in Parchim on 9 February 1945. It was now decided to form two "He 162 Interceptor *Staffeln*" from the remnants of the two *Gruppen*, 1./JG 1 in Marienehe and Bernburg and 2./JG 1 in Heidfeld near Vienna. On 24 February *2. Staffel* under the command of Lt. Hachtel transferred to Heidfeld near Schwechat (Vienna). Two days later the OKL ordered 3./JG 1 to Parchim. Since there were no He 162s available there, most of the *Staffel* flew on to Vienna, where a single He 162, the M19, was available for training purposes. When no further He 162s became available, part of 2. and 3./JG 1 returned to Parchim.

Eight of I./JG 1's pilots succeeded in converting to the *"Volksjäger"* by 7 March 1945. Because of the temperamental nature of the BMW 003 turbojets, during training the pilots were forbidden to exceed 500 kph or remain in the air for more than 15 minutes.

The general situation at Heidfeld was more than difficult, as there were still insufficient trained ground crews. The Heinkel personnel more than had their hands full keeping their machines serviceable. Then, on 14 March, Uffz. Daus (or Tautz) was killed in the He 162 M19. The requested

The aircraft of Oblt. Emil Demuth, 3./JG 1, at Leck 1945.

Pilots of JG 1 in front of their aircraft. From left to right: Hptm. Künnecke, Oblt. Demuth and Major Zober.

Oblt. Demuth and Major Zober in front of Lt. Schmitt's He 162. Photo: Archiv Peter Petrick.

conversion of the BMW 003 to J2 fuel (kerosene), a result of unavailability of B4 fuel, proved much more difficult than anticipated.

The plans which reached the *Geschwader* on 20 March 1945 gave no hint of all these problems. The conversion of JG 1 requested by the GenQu was in full swing, at least on paper. By May 1945 the *Geschwader* was supposed to consist of the *Stabsstaffel*, with 16 He 162s and three *Gruppen* with 52 aircraft each. The reality was much worse. The Heidfeld detachment was supposed to receive production aircraft 1 to 5, while machines 6 to 13 had been allocated for testing at Lechfeld and 14 to 20 for trials at Roggenthin near Rechlin. In spite of everything, on 21 March 1945 the interceptor *Staffel* 2./JG 1 did not have a single He 162.

On 26 March 1945 all of the plans were revised again. The first 18 He 162s were now to go to the Heidfeld detachment, and only then would tactical trials begin. On 30 March, however, there was not a single He 162 in Lechfeld or in Rechlin. The first two machines arrived there in mid-April 1945 and bore the codes E3 + 51 and E3 + 52. An emergency test program began soon afterwards, however, it was brought to a quick end by the Allied advance.

At the end of March the last ten flyable He 162s, including several prototypes, flew from Vienna to Langenlebarn and Hörsching. The last casualty suffered during testing was company pilot Wolfgang Lüddemann, who crashed on 6 April. Soon afterwards a number of the *"Volksjäger"* flew via Hörsching, where several were left behind unserviceable, and Munich to Lechfeld. On 31 March Heinkel company pilot Huldreich Kemnitz crashed during the ferry operation and lost his life. Only three of the ten He 162s arrived on schedule. Following the Allied advance into this area, the He 162 M20 and M23 were evacuated to Munich-Riem.

Not until 1 April 1945 did the first four He 162s reach the Heidfeld interceptor *Staffel*. Two more followed by 7 April. With serviceability averaging two to three machines at best, even daily training flights were not possible.

Close-up of the nose of a He 162 of I./JG 1. Photo: Archiv Manfred Griehl.

From the interceptor *Staffel* was formed the Stabsstaffel JG 1. At the same time I./JG 1 was ordered to transfer to Ludwigslust and *II. Gruppe* to Garz (instead of Warnemünde). *I. Gruppe* received its first He 162 on 8 April 1945. The aircraft were supposed to be taken on strength by *III. Gruppe*, however the *"Volksjäger"* became part of I./JG 1. Conversion of pilots was completed by 11 April, with as many as 16 machines involved in training. The fuel situation permitted about ten to twelve practice flights per day, and soon 30 to 40 pilots had gained their first experience in the *"Volksjäger."* The *Gruppe* lost *Oberfähnrich* Stenschke and *Unteroffiziere* Enderle and Werner during training. In spite of all efforts, II./JG 1 still did not have a single jet fighter. In fact, the *Gruppe* did not commence conversion training until 20 April 1945.

On 14 April I./JG 1 under the command of Oblt. Demuth moved from Ludwigslust to Husum. During the transfer flight Lt. Rudolf Schmitt made contact with a British Spitfire. No victory can be confirmed.

Two days later the *Gruppe* had to move from Husum back to Leck. Enemy fighters were encountered at about 1530 hours. The formation of eight to ten He 162s was able to exploit the superior speed of its machines and avoid the threat. On 18 April the majority of the ground personnel arrived in Leck from Warnemünde.

In spite of the improvements that had been made, the *"Volksjäger"* still presented problems to its pilots. On 20 April 1945 Lt. Schmitt was forced to use his ejection seat. Three days later Uffz. Steeb of 1./JG 1 bailed out from a height of 300 meters. On 24 April the *Kommandeur* of II./JG 1 was

Cockpit of a He 162. On the right is the rigidly-mounted flare pistol. Photo: Archiv Manfred Griehl.

fatally injured when he ejected from his aircraft without first jettisoning the canopy. During the period 20-26 April 1945 Fhr. Halmel, Uffz. Fendler, and Uffz. Rechenbach were all killed in flying accidents. They are buried in the cemetery in Leck. The fuel shortage further reduced the number of serviceable machines. Hptm. Ludewig and his wingman, Fw. Gehrlein, were obliged to make forced landings after running out of fuel.

Beginning on 25 April 1945 the He 162s were employed against tactical aircraft in the Flensburg—Heide—Schleswig area. A pair of aircraft belonging to *I. Gruppe* attempted to intercept a Mosquito but returned empty handed. One day later, on 26 April, Lt. Hanf intercepted a Typhoon but failed to shoot it down. By 29 April Gerhard Hanf took part in six more scrambles in the *"Volksjäger."* Lt. Schmitt's log book reveals that on 4 May 1945 he engaged a Typhoon in his aircraft "White 1," scoring effective hits. The victory was not officially confirmed.

On 27 April the remaining elements of II./JG 1 arrived in Leck from Mecklenburg, having stopped in Kaltenkirchen to refuel. Deliveries of new machines gradually came to a halt, and conversion training and flying activity soon ended.

On 2 May 1945 *Oberstleutnant* Ihlefeld also arrived in Leck with the *Geschwaderstab*. One day later the unit's command staff consisted of:

Kommodore	*Oberstleutnant* Ihlefeld
Kommandeur I./JG 1	*Major* Zober
1. Staffel JG 1	*Hauptmann* Künecke
2. Staffel JG 1	*Hauptmann* Ludewig
3. Staffel JG 1	*Oberleutnant* Demuth
Kommandeur II./JG 1	*Hauptmann* Rahe
4. Staffel JG 1	*Hauptmann* Gallowitsch
5. Staffel JG 1	*Hauptmann* Bergholz
6. Staffel JG 1	*Hauptmann* Zipprecht

One day later it became clear that this elaborate organization could no longer be maintained. I. and II./JG 1 had to be combined as I. (*Einsatzgruppe*)/JG 1 under the command of Herbert Ihlefeld. The *Gruppe* consisted of *1.* and *2. Einsatzstaffel* (*Major* Zober and *Hauptmann* Ludewig).

There were several flying accidents shortly before the end of the war. Fw. Köhler crashed on the runway at Leck, wrecking his He 162 A-2. Oskar Köhler was pulled from the wreckage by *Oberleutnant* Demuth.

Missions continued to be flown over Schleswig-Holstein against Allied tactical aircraft during the first days of May. The pilot of the British aircraft claimed shot down by Lt. Schmitt stated that he had been brought down by anti-aircraft fire. In any case, he spent the last days of the war in JG 1's mess waiting for his comrades, who were now not far away.

The *Einsatzgruppe* (operational *Gruppe*) was ordered to make one final transfer flight, to northern Denmark or southern Norway, however, the fuel shortage made this impossible.

On 5 May 1945 demolition charges were installed in the operational *Gruppe*'s thirty or so remaining He 162s, with the intention of blowing them up before Allied troops arrived. At about 2400 hours, however, the *Kommodore* ordered the charges removed.

The next day British troops reached the airfield and discovered the jet fighters. One of the English sentries decided to demonstrate the He 162's ejection seat to his comrades. He climbed into the cockpit and activated the ejection seat. He did not survive the demonstration, as he landed on his head.

The remaining elements of JG 1 stayed on the airfield until 15 May 1945. On 21 May they were moved to quarters in Schnörholm and finally to Hennstedt.

An order dated 26 February 1945 stated that JG 400 was to be the second unit to equip with the *"Volksjäger."* The unit's *I.* and *II. Gruppe* would hand in their Me 163s prior to receiving the He 162. The general collapse prevented this plan from being carried out, however.

The first of an initial batch of five prototypes of the two-seat trainer version of the He 162 was completed in March 1945. This unpowered version was test-flown by the Reich Gliding School in Trebbin/ Brandenburg.

The He 162 M1 prior to its first flight on 6 December 1944. In the cockpit is Dipl.-Ing. Gotthold Peter. Peter was to die four days later in the crash of the same machine.

This jet engine and cockpit section was used to train pilots and mechanics how to operate the new turbojet.

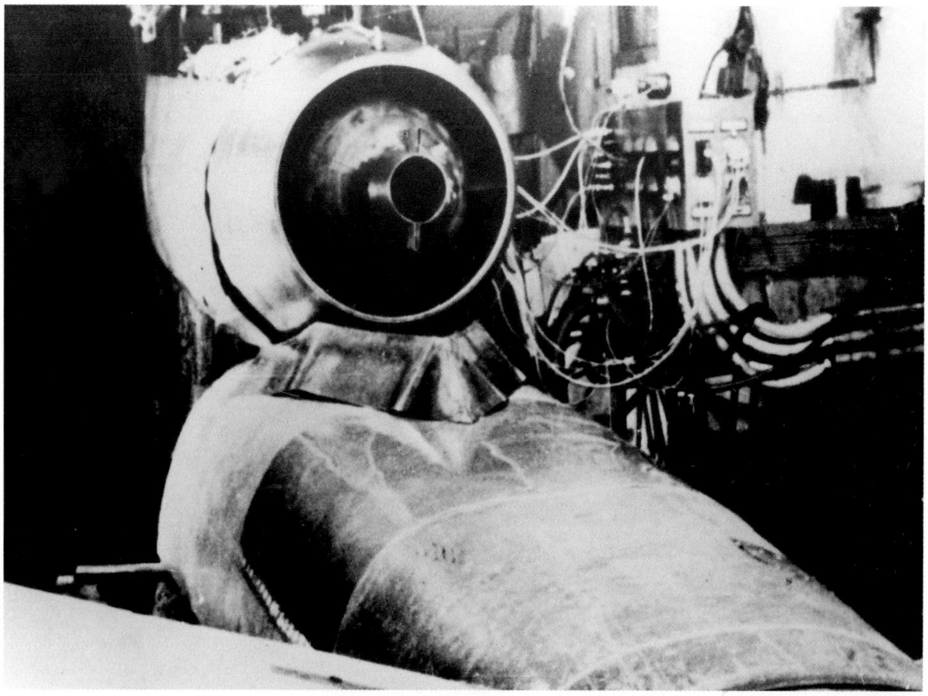

Testing a BMW 003 turbojet engine in the BMW factory in Allach near Munich.

The various production departments were separated by wooden walls.

After the war the Seegrotte site was emptied under Soviet supervision. The bulk of the fuselages and other components found there were scrapped.

Entrance to the underground production site in Mödling near Vienna. The Seegrotte factory can be visited today, offering a good impression of the extent of the He 162 production program, which occupied two levels.

View of the elevator shaft between the first and second levels. The sliding stage was capable of lowering a complete He 162 fuselage to the exit level.

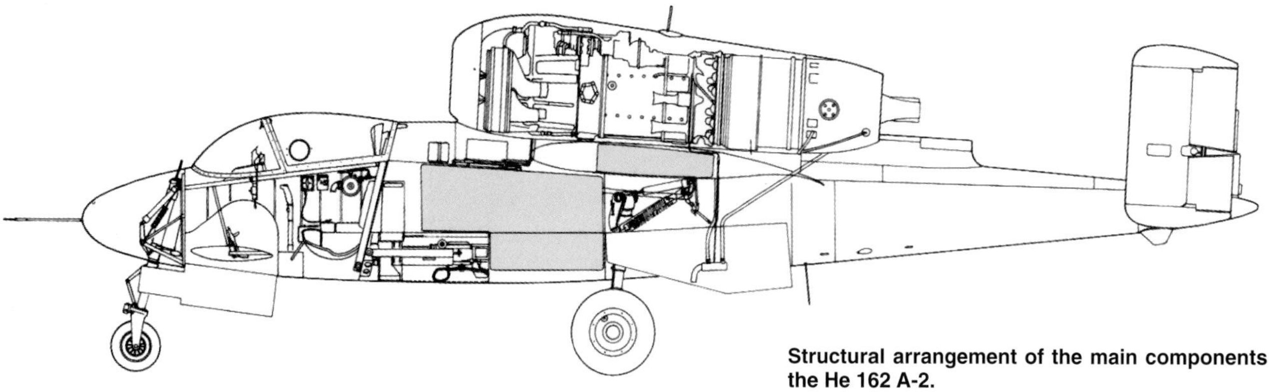

Structural arrangement of the main components of the He 162 A-2.

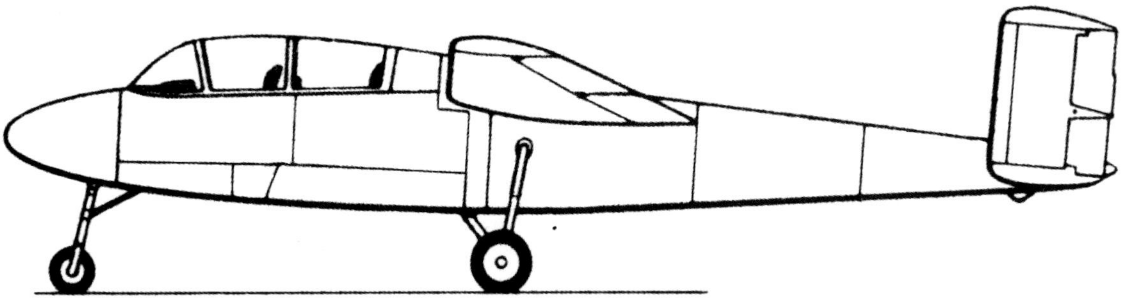

Side-view of the He 162 S.

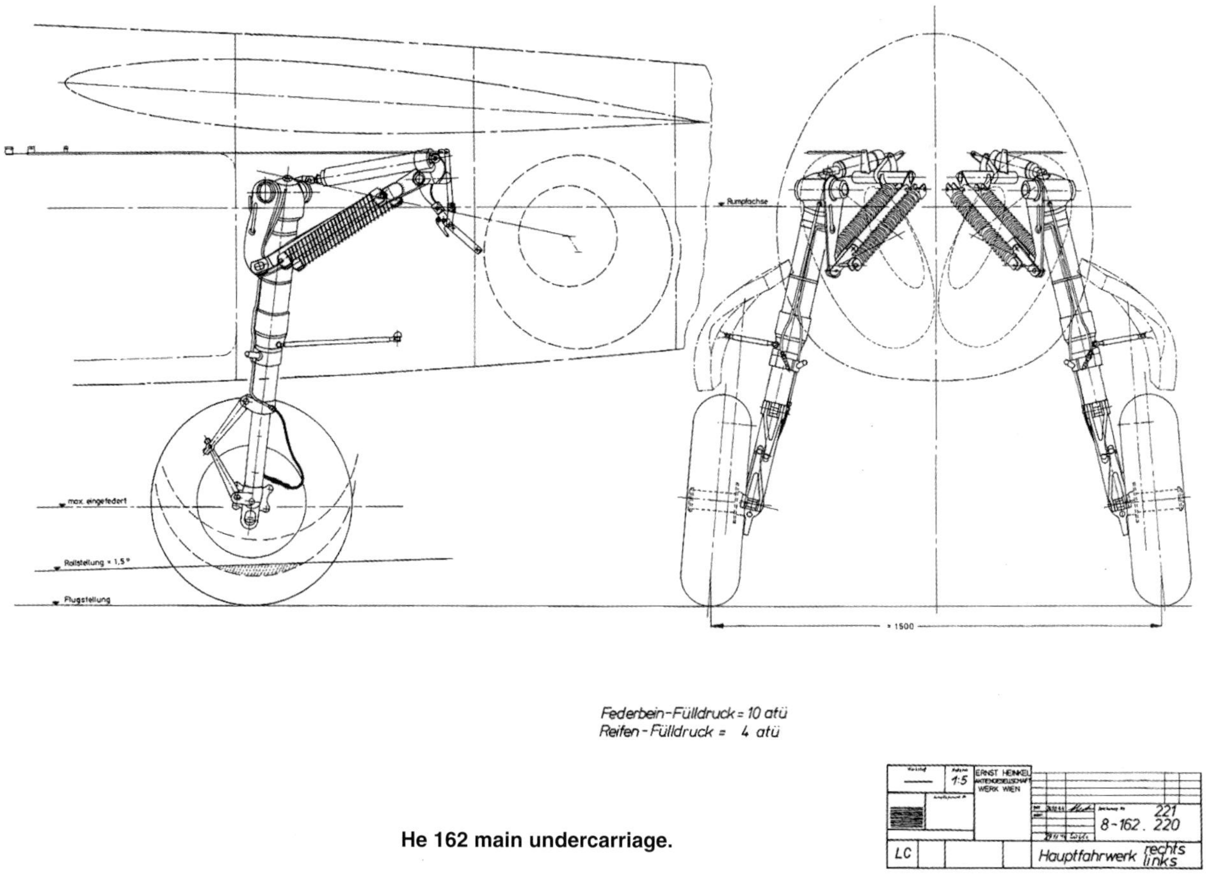

He 162 main undercarriage.

"Nervenklau"
1
120077

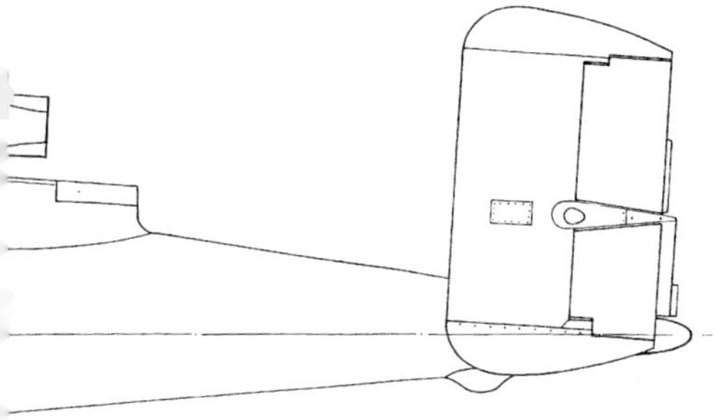

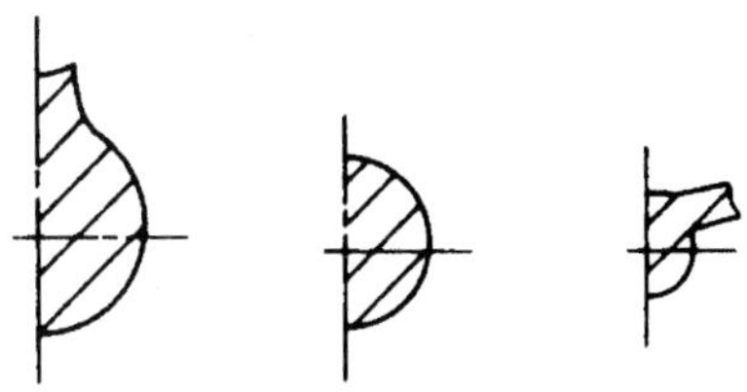

G H J

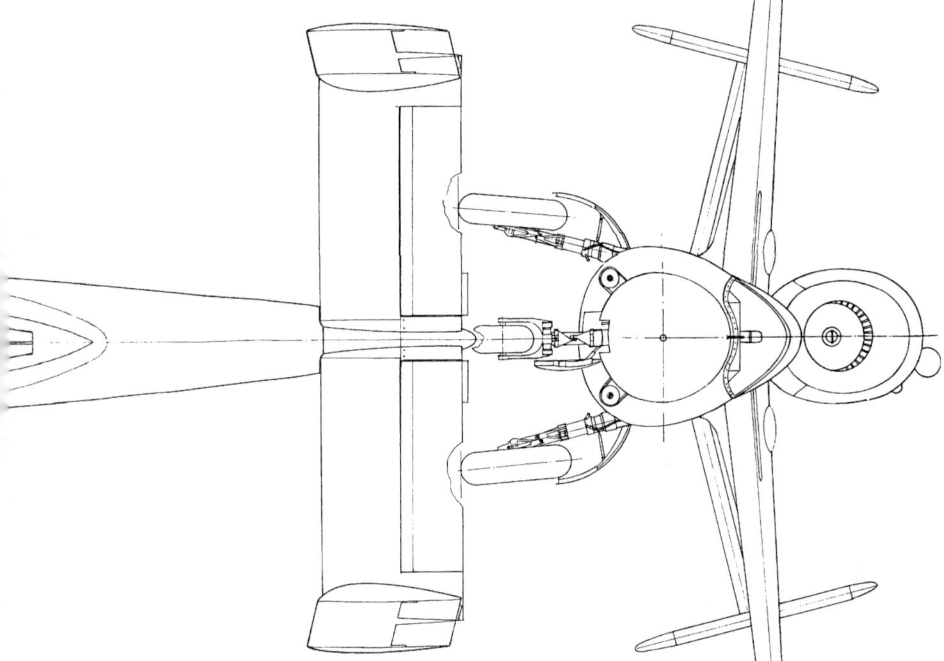

Three-view drawing: He 162 A-2, Werk-Nr. 120077, Red 1 "Nervenklau," 2./JG 1, May 1945.

Drawings: Manfred Meyer

He 162 A-1/M6, Werk-Nr. 200006, VI + IF, maiden flight 23 January 1945, crashed on 4 February as a result of a jammed rudder.

He 162 A-2, Werk-Nr. 120028, White 3, I./JG 1, May 1945, note upper surfaces in Green 83 and old-style cross.

He 162 A-2, Werk-Nr. 120074, Yellow 11, aircraft of Oblt. Emil Demuth, Staffelkapitän of 3./JG 1, May 1945. The nose colors black-white-red indicate that the pilot wore the German Cross in Gold. The sixteen victory bars on the vertical fin represent victories scored while flying other types.

He 162 A-2, Werk-Nr. 120231, White 6 of 1./JG 1. The forward fuselage is unpainted, although the seams have been filled.

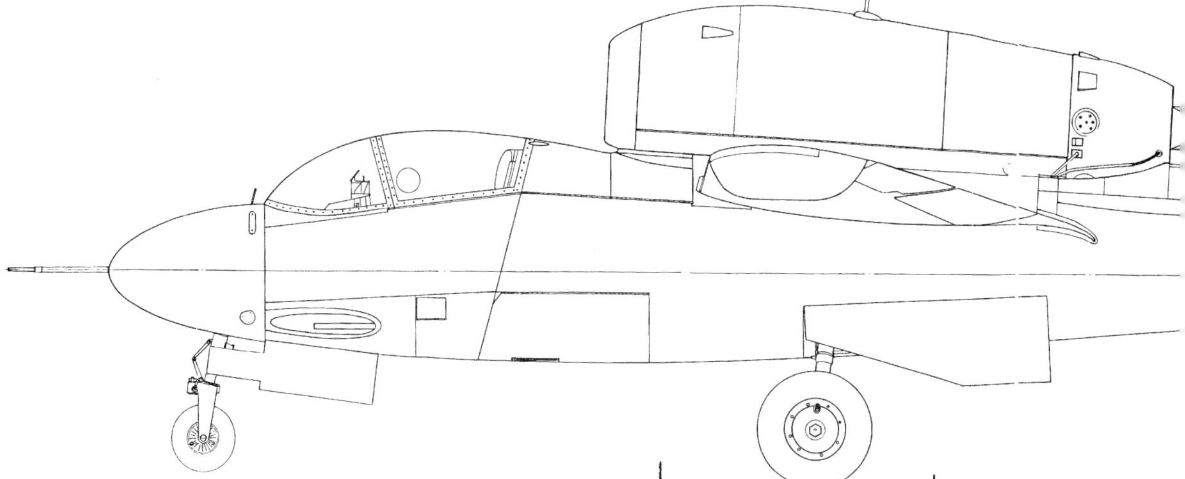

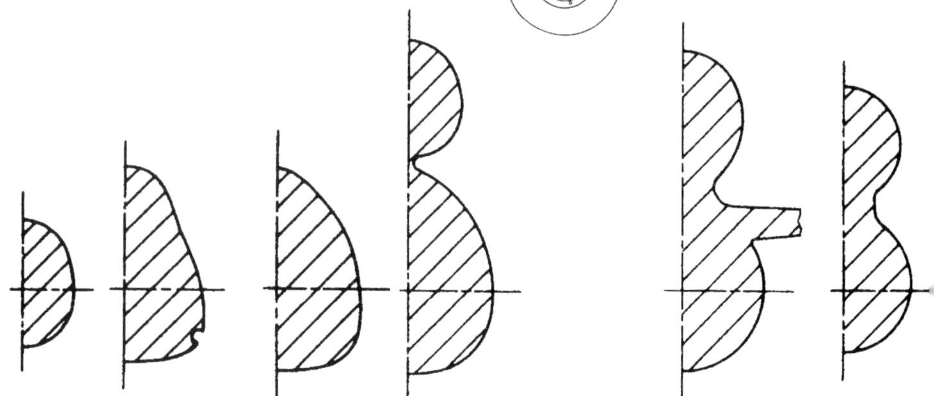
A
B
C
D
E
F

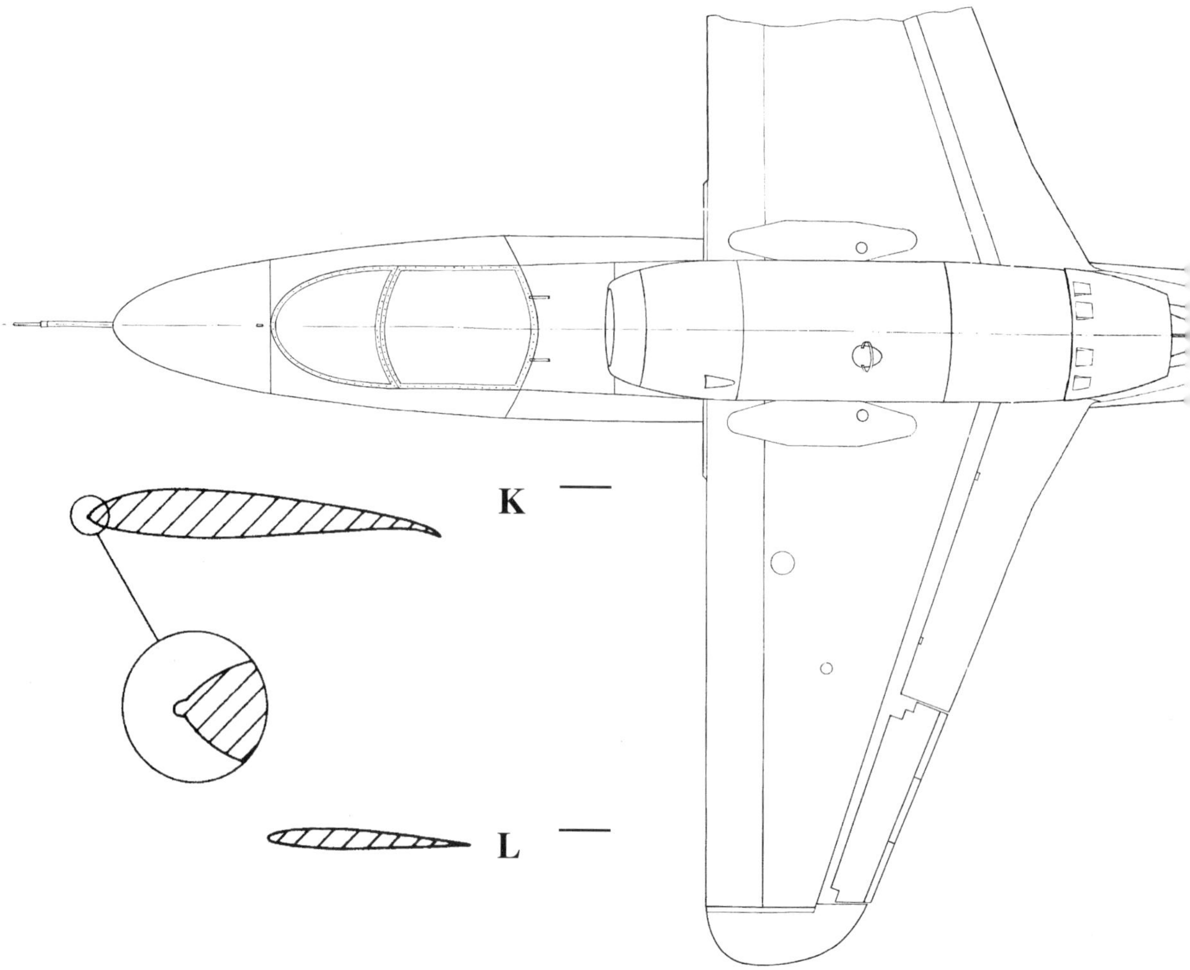
K
L

This photo, taken during flight operations in April 1945, depicts aircraft "White 5." Fw. Enderle was killed in the crash of this machine on 13 March 1945.

Servicing a He 162 A-1 of I./JG 1 "Oesau" on an airfield in northern Germany.

JG 1 was the Luftwaffe's sole Jagdgeschwader to be equipped with the "Volksjäger." A total of 50 He 162 A-1s and A-2s was delivered to the unit and to the E-Stelle Rechlin.

Contrary to orders, the He 162s at Leck were not blown up. Instead, most fell intact into the hands of the British soldiers, which arrived just prior to the end of the war.

Another view of I./JG 1's He 162s at Leck. The aircraft indicated by the cross was flown by Lt. Gerhard Hanf.

Mechanics work on "White 3," a He 162 A-2. Following minor repairs to the engine, the aircraft was cleared to fly on 15 March 1945.

The He 162 M23 was test-flown by Gerhard Gleuwitz on 19 March 1945, and on 1 May was in a hangar on Munich-Riem airfield.

The He 162 M20 and the M23 in Munich-Riem. Both machines had sustained considerable damage.

Side view of the He 162 M20. It was damaged in a crash on 24 February 1945, and in April was found by the Americans at Munich-Riem.

The He 162 V23, VI + IP, was towed to a parking area for examination. The engine has already been removed.

Extract from the technical directives for the loading and transport of the He 162.

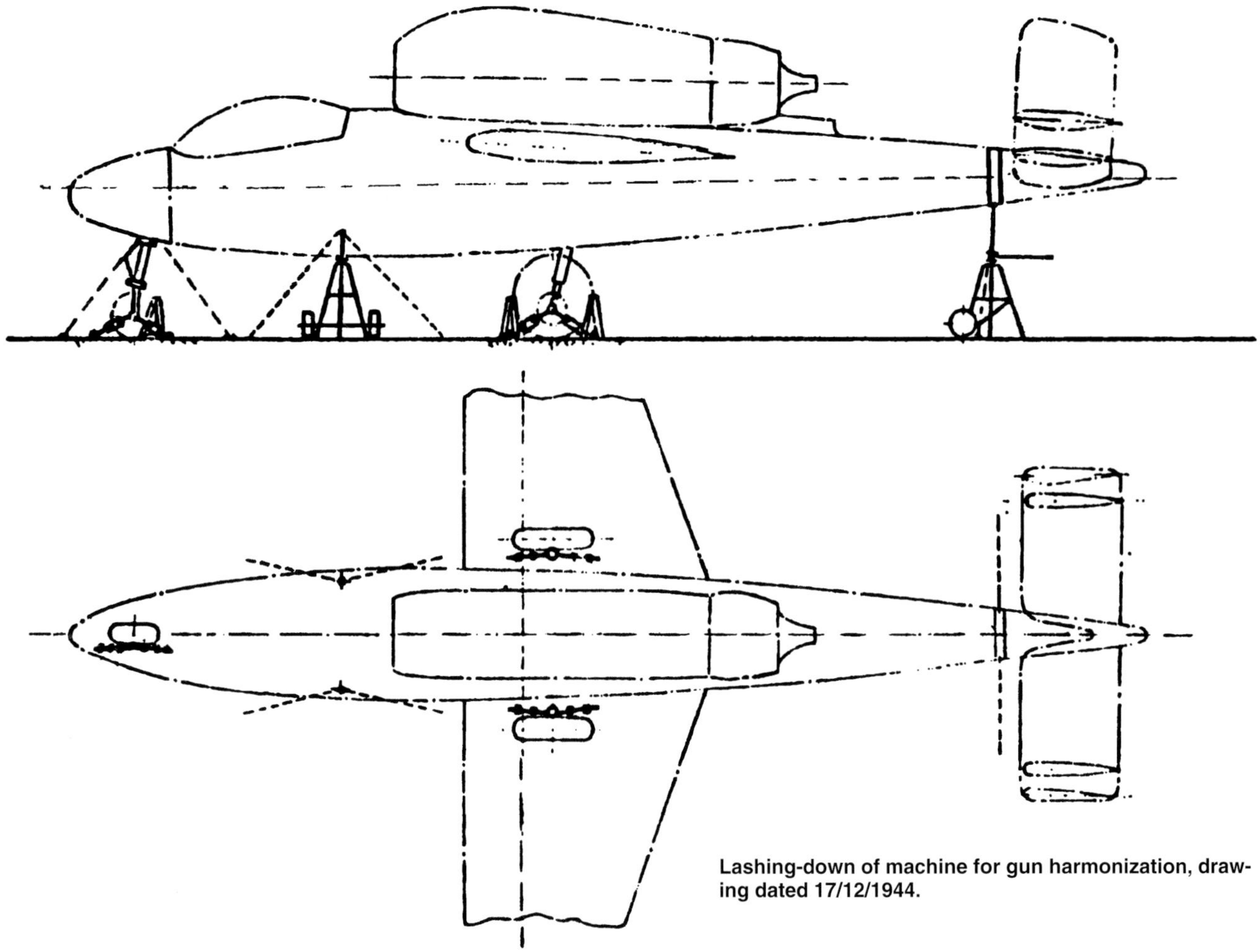

Lashing-down of machine for gun harmonization, drawing dated 17/12/1944.

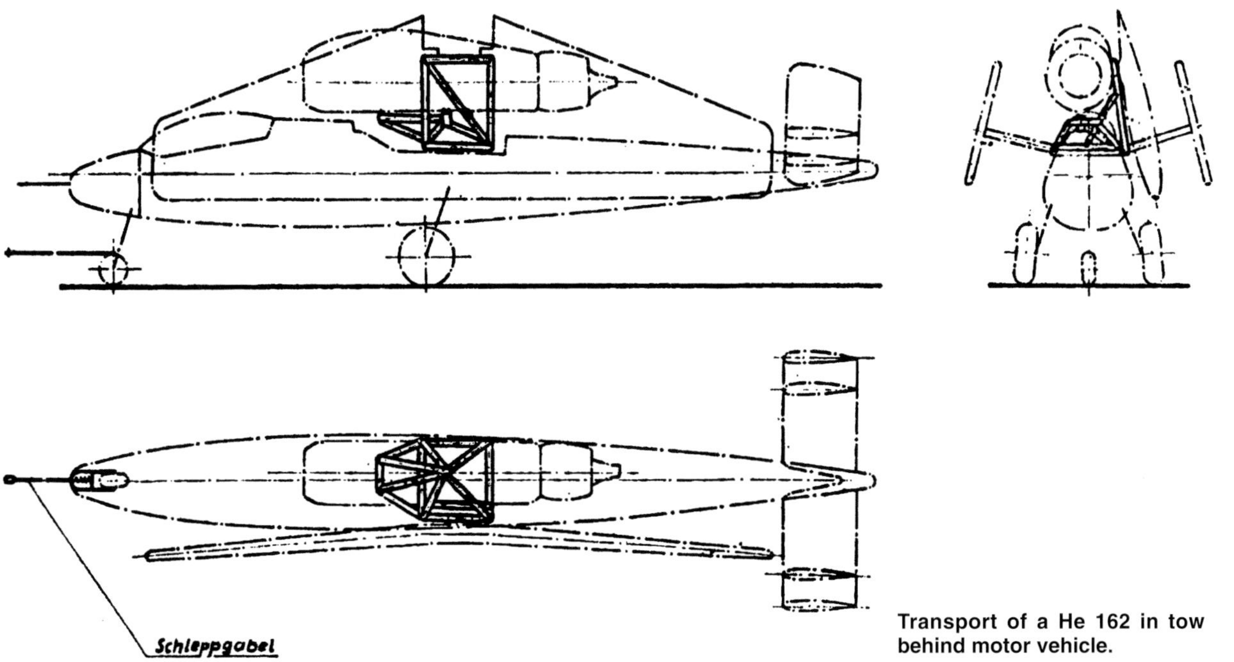

Transport of a He 162 in tow behind motor vehicle.

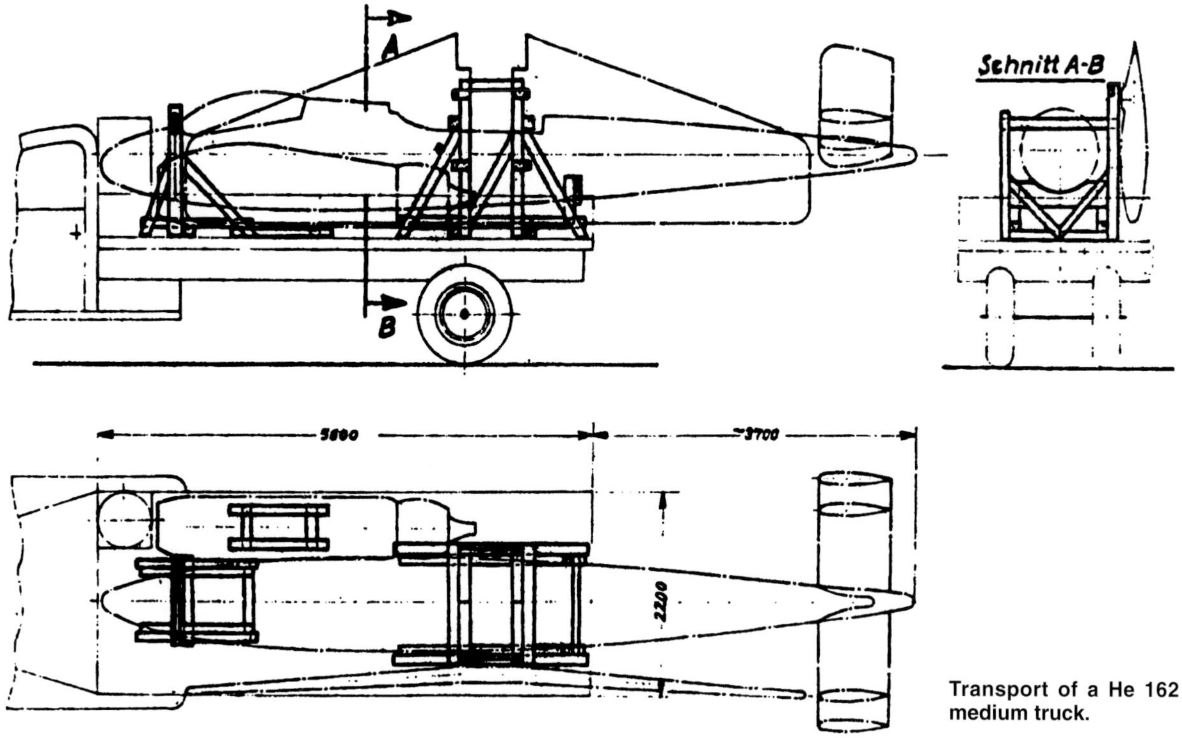

Transport of a He 162 on a medium truck.

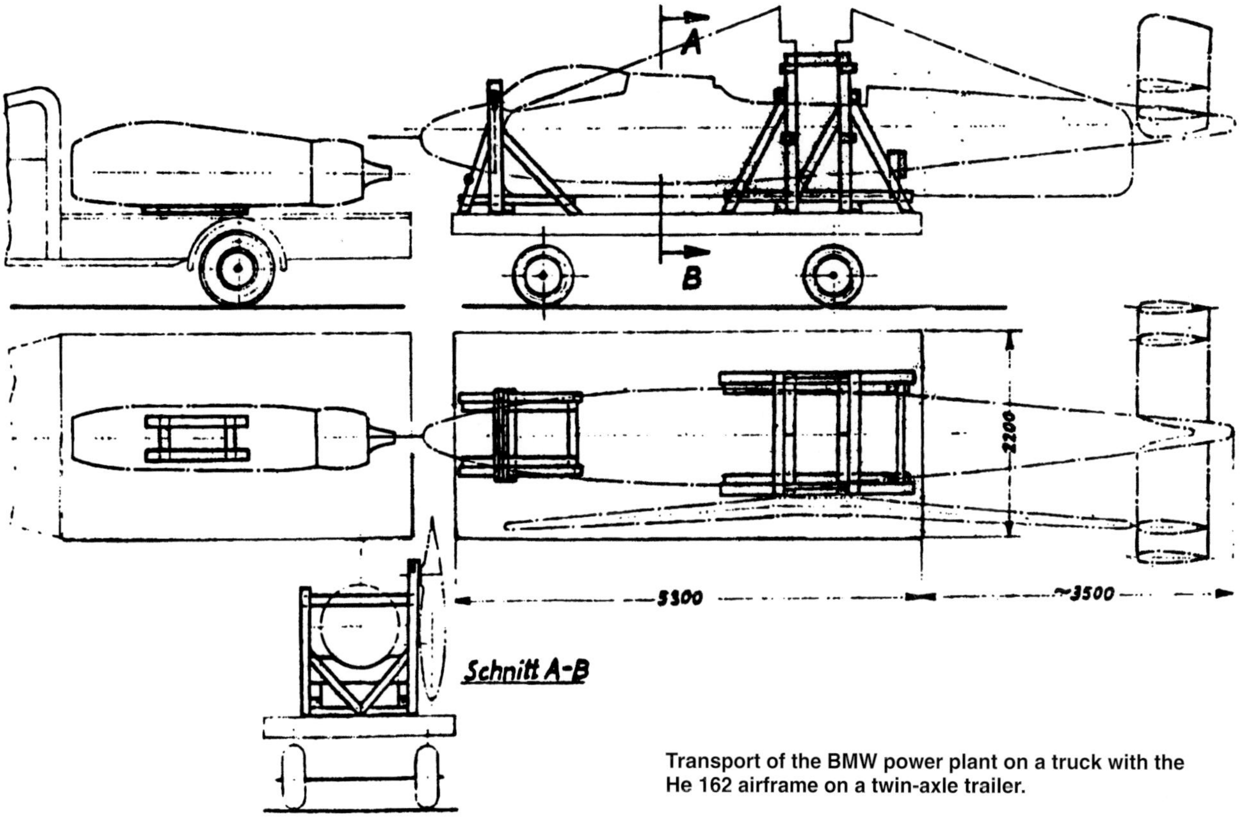

Transport of the BMW power plant on a truck with the He 162 airframe on a twin-axle trailer.

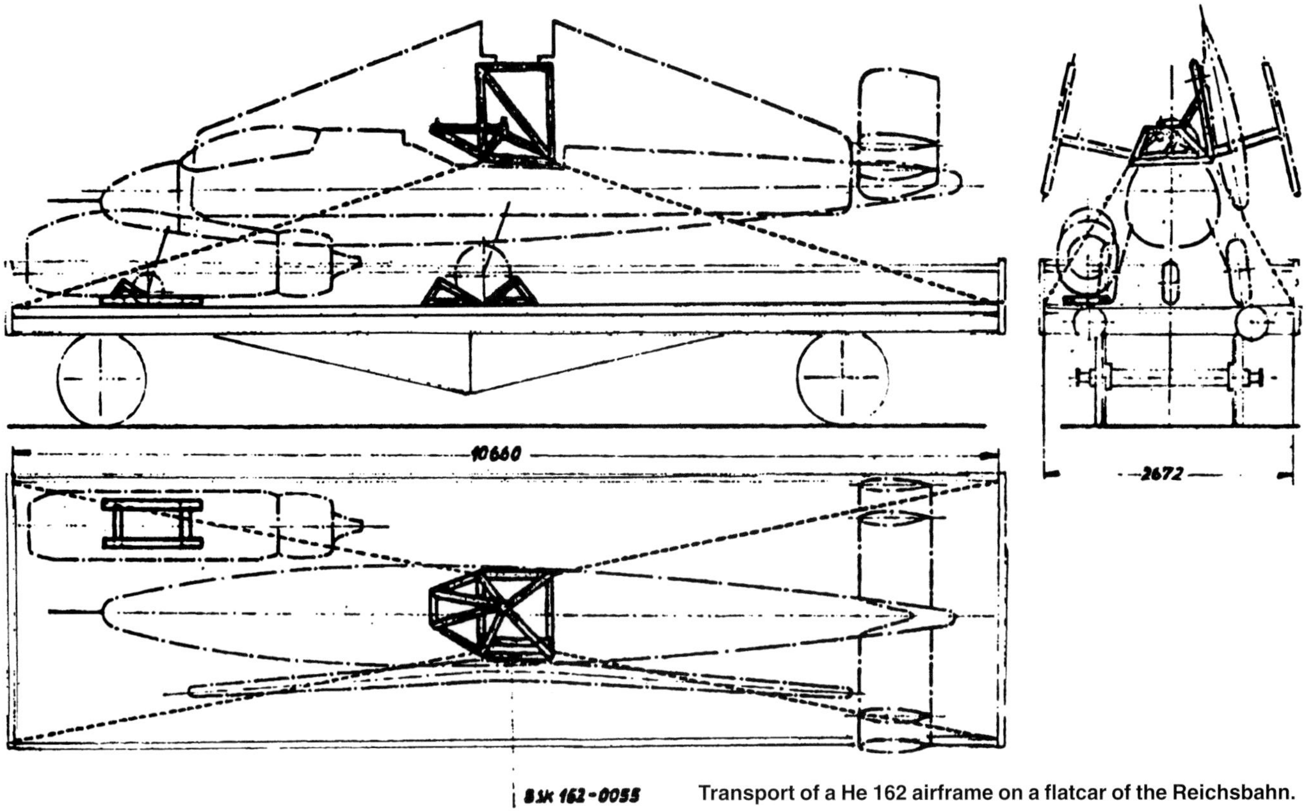

Transport of a He 162 airframe on a flatcar of the Reichsbahn.

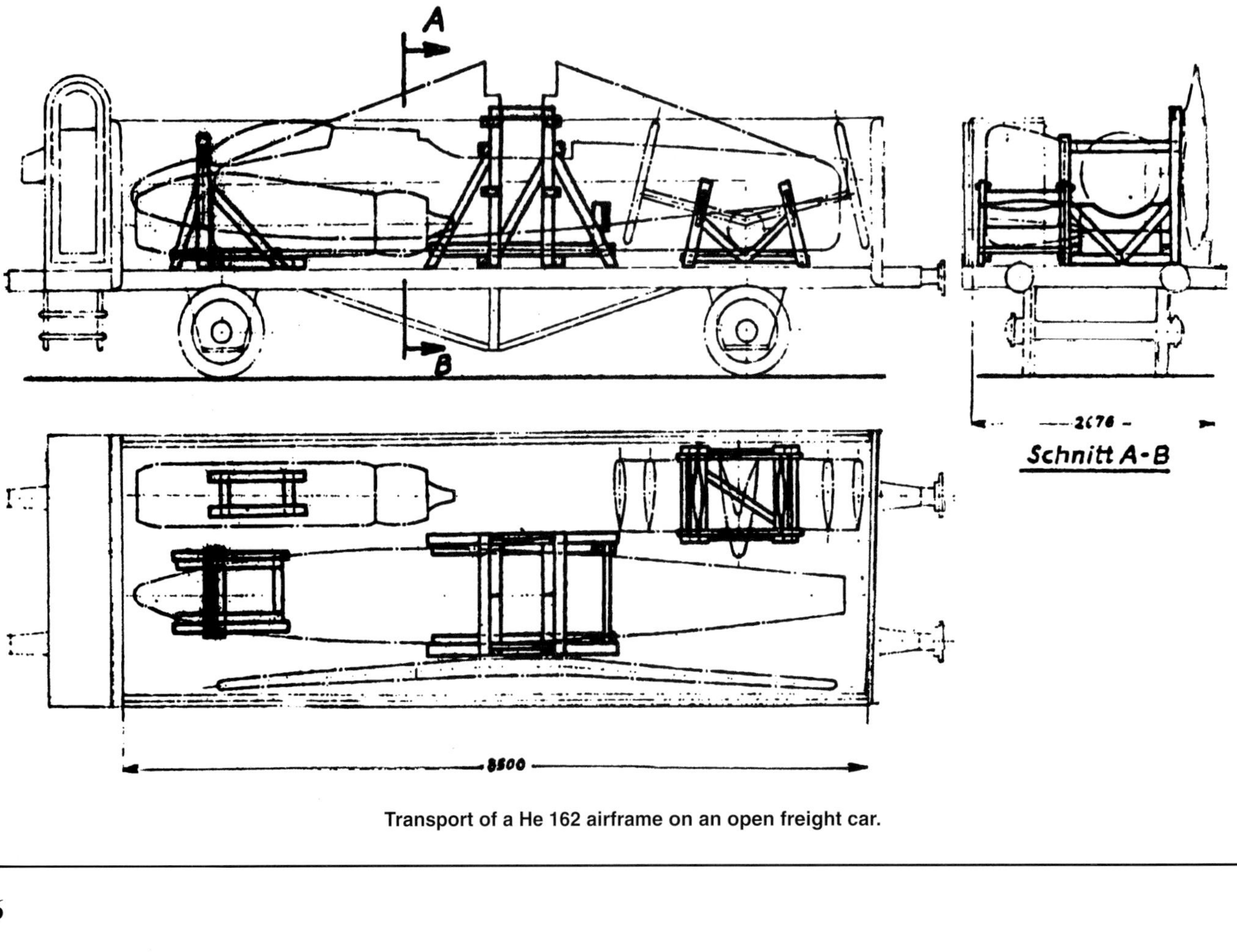

Transport of a He 162 airframe on an open freight car.

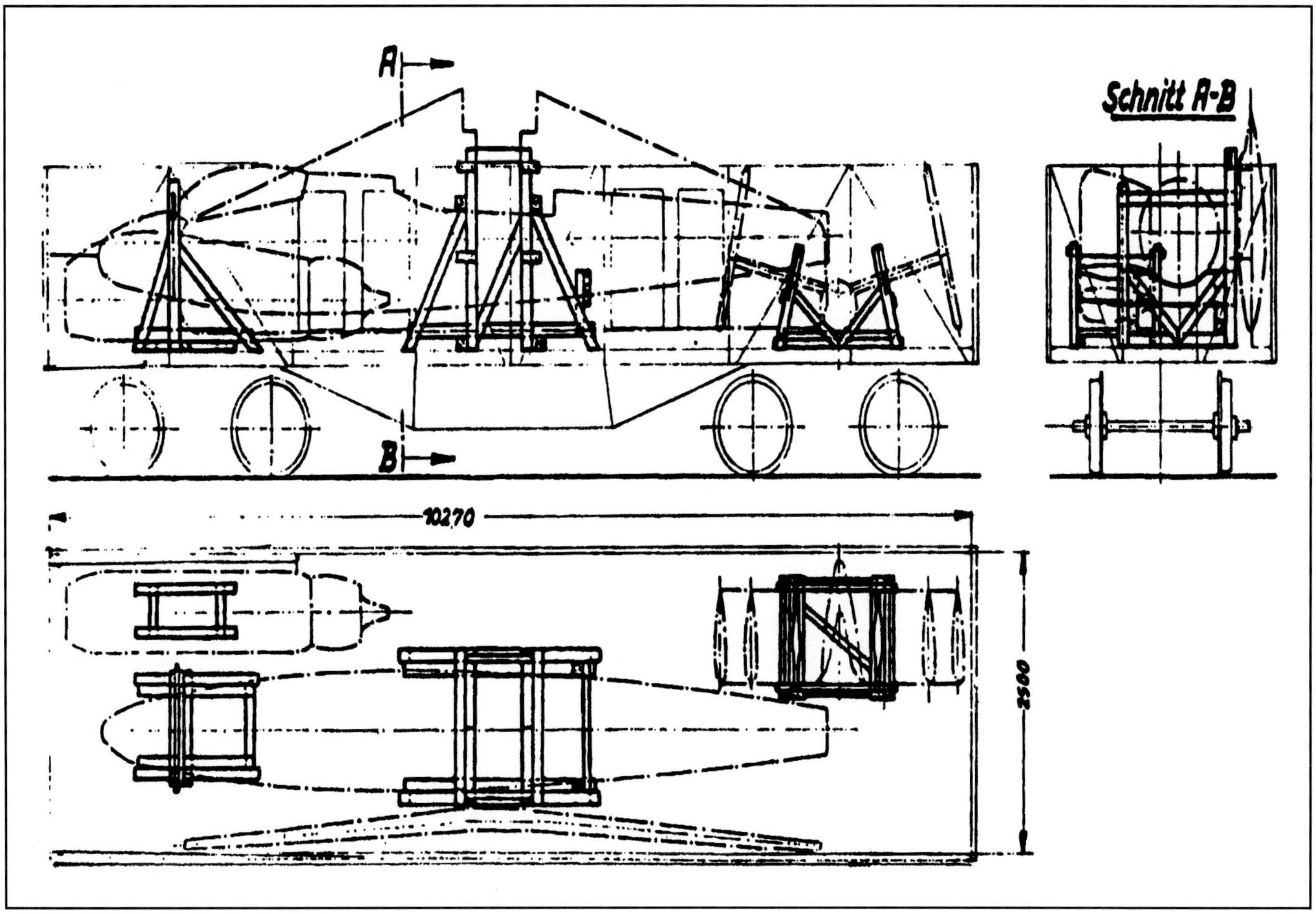

Above and below: Transport of a He 162 airframe on a four-axle freight car.

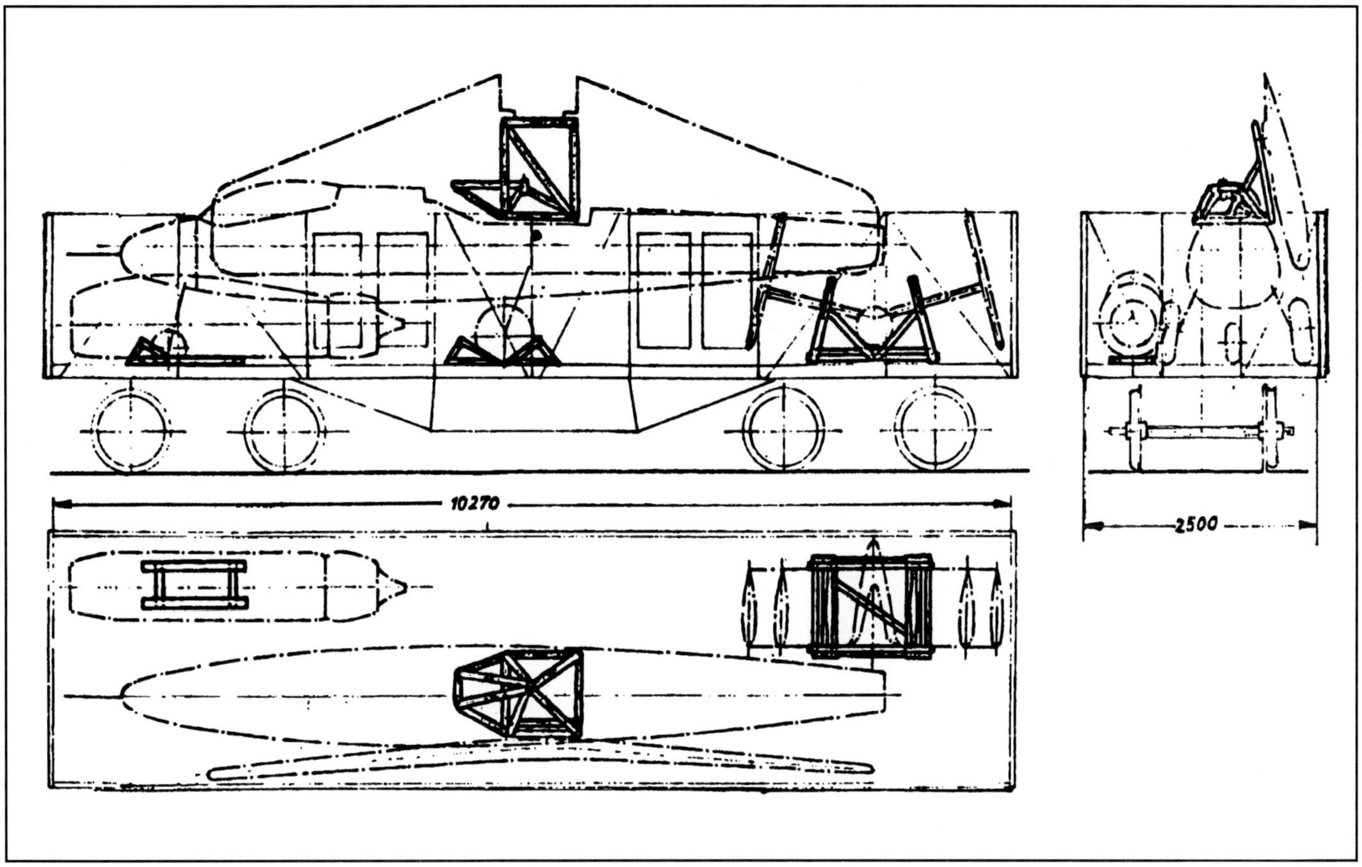

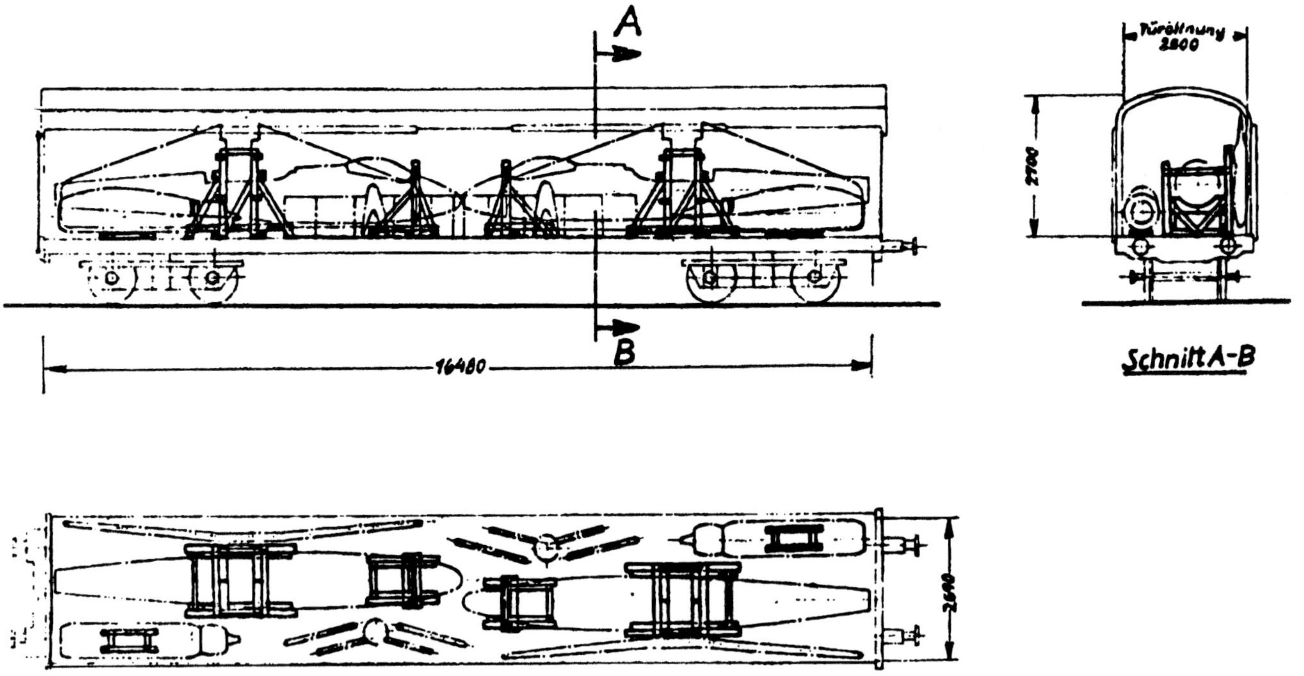

Transport of two He 162s in a covered, four-axle freight car.

Captured Heinkels

After the war this He 162 was tested by the French at their test station in Istres.

Heinkel He 162 A-1 in Leck. For cg reasons, the aircraft's tail had to be supported with the cannon removed.

The following sequence of photos depicts two Heinkel He 162s in Soviet service wearing the tactical numbers 01 and 02.

The emblem on the vertical fin of the He 162 is the emblem of the L II test center.

At least two He 162 *"Volksjäger"* were tested at the N II VVS and L II test centers.

Heinkel He 162 A-2 at the British Farnborough test station. This aircraft, which is registered AIR MIN 61 (Werk-Nr. 120072), was supposed to undergo extensive testing, however, it was lost in a crash near Aldershot on 9 November 1945. The aircraft's pilot, Flight-Lieutenant Marks, was killed.

The Heinkel He 162 M27 was discovered by American troops in Staßfurt in damaged condition.

Heinkel Projects

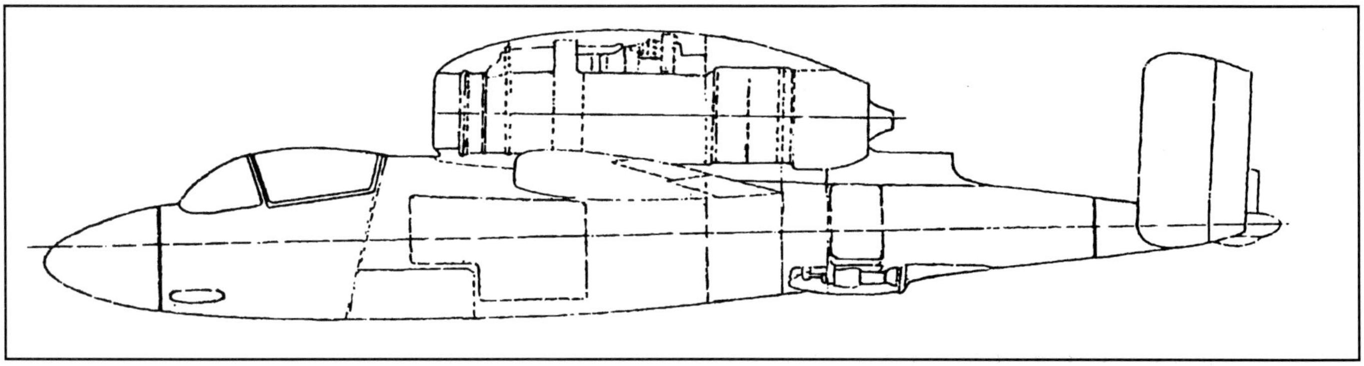

Original drawing of a He 162 with combined propulsion system consisting of a turbojet engine and rocket booster engine.

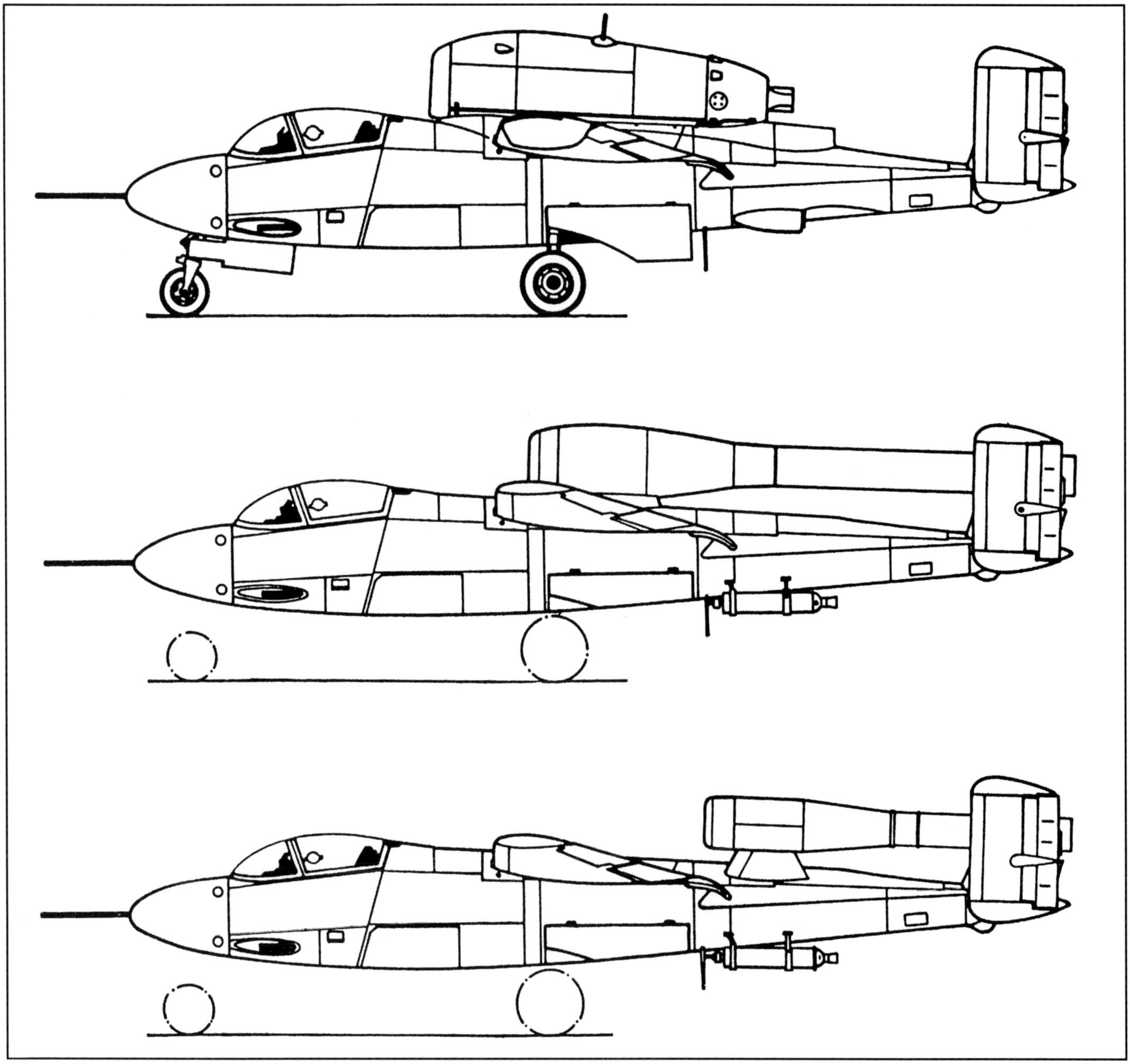

Three planned versions based on the He 162 A airframe (from top to bottom): high-altitude fighter with mixed turbojet-rocket propulsion, fighter aircraft with two Argus As 014 pulse-jet engines, fighter aircraft with one Argus As 044 pulse-jet.

Mistel Project

Projected Mistel combination consisting of a He 162 A as control aircraft and a flying bomb powered by two BMW turbojet engines (derived from the Ar E-377).

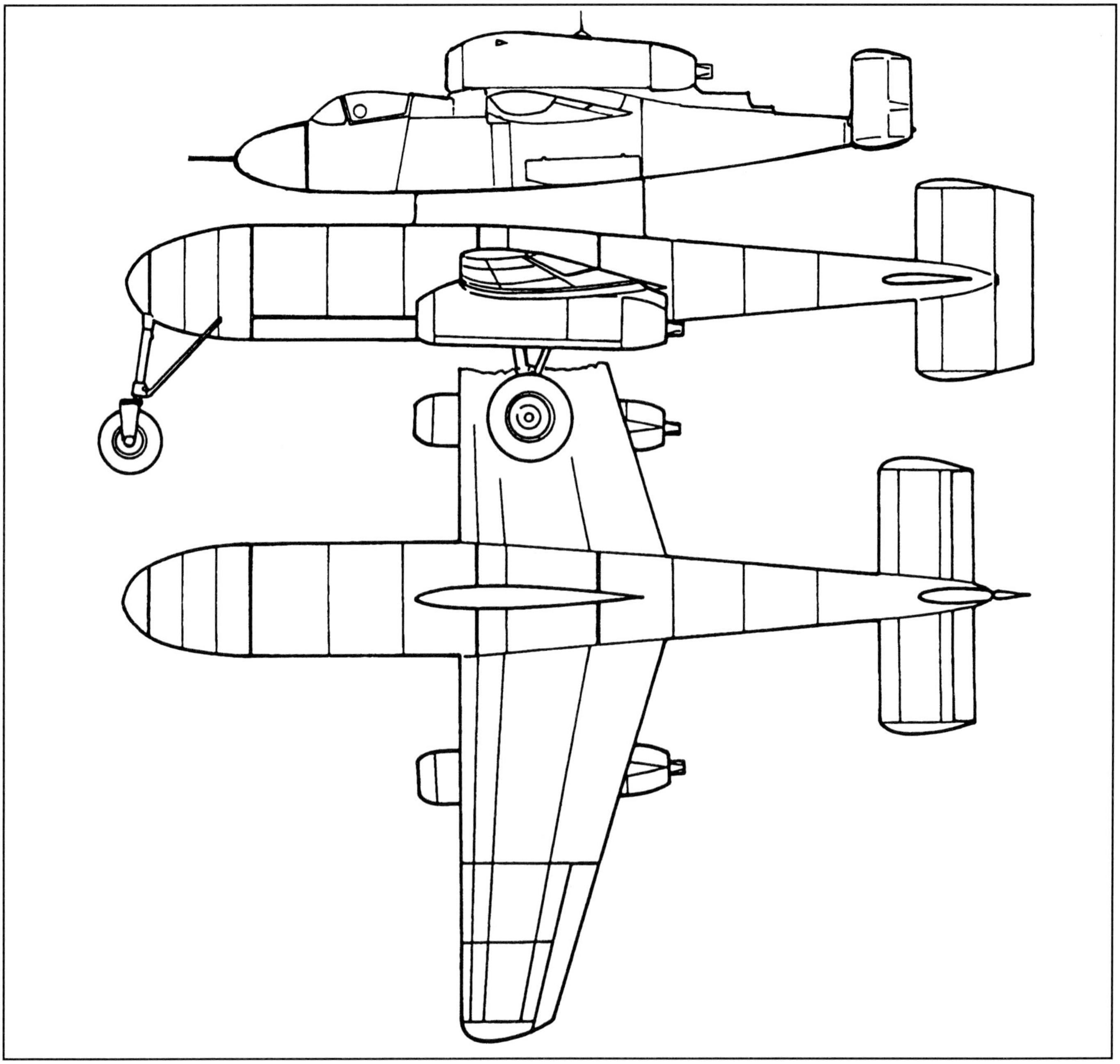

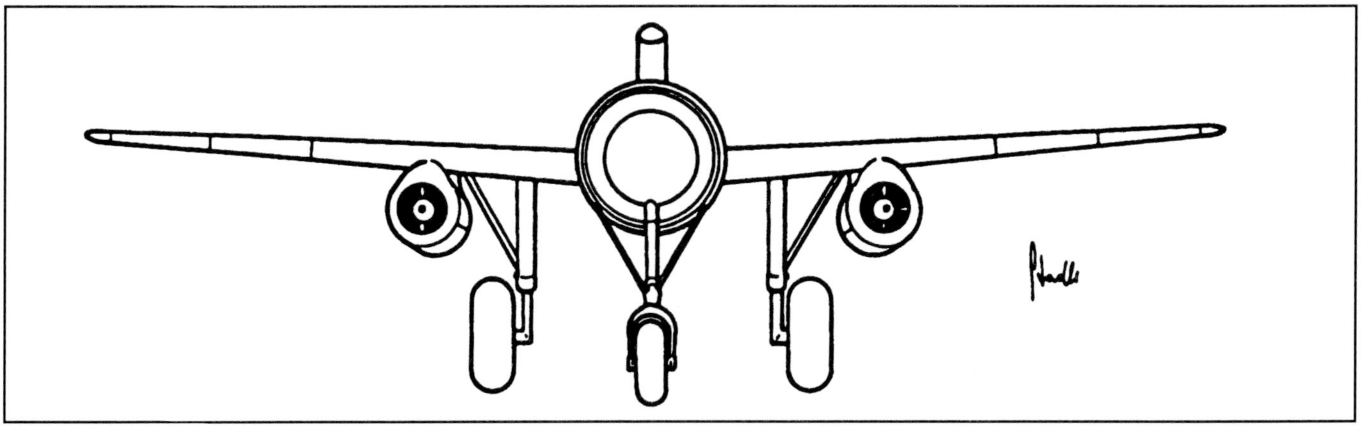

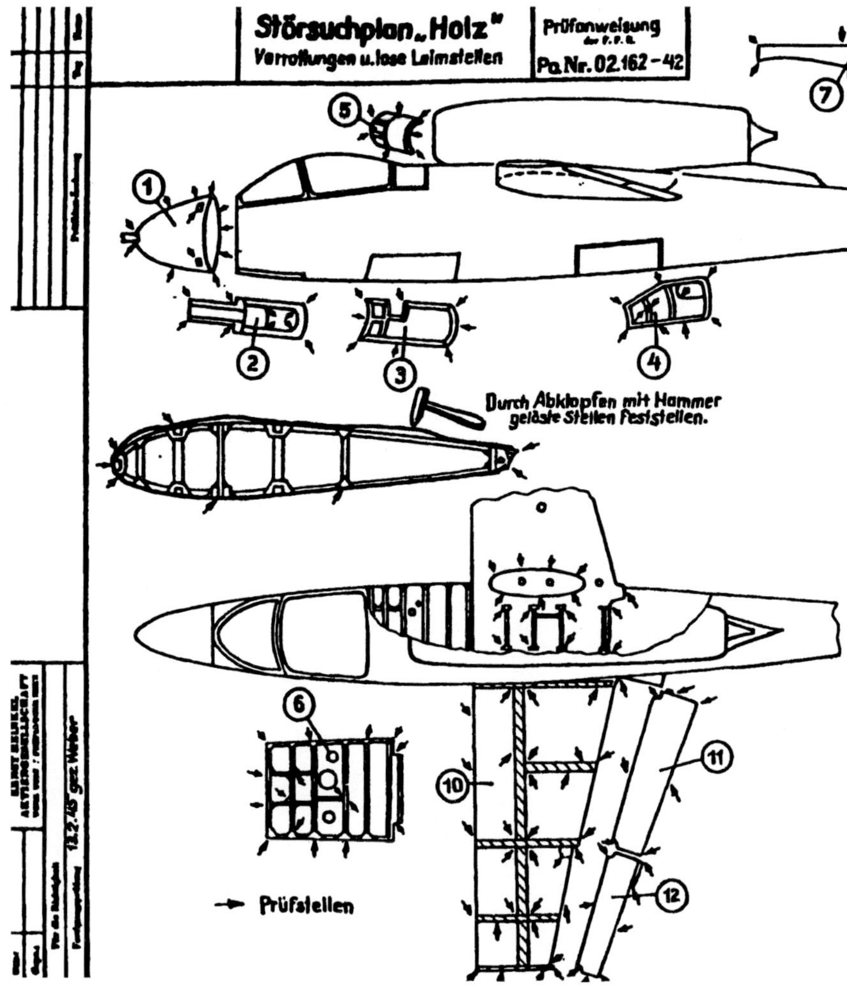

Fault location guide for the wooden components of the He 162 A-1 and A-2 fighter aircraft with numbered fault search sequence:
1: nose cone;
2: nosewheel door;
3: cannon access panels, left and right;
4: undercarriage doors, left and right;
5: ammunition tank cover;
6: fuel tank support frame;
7: split engine cowling, left and right;
8: vertical stabilizers, left and right;
9: rudders, left and right;
10: wing;
11: landing flaps, left and right;
12: ailerons, left and right.

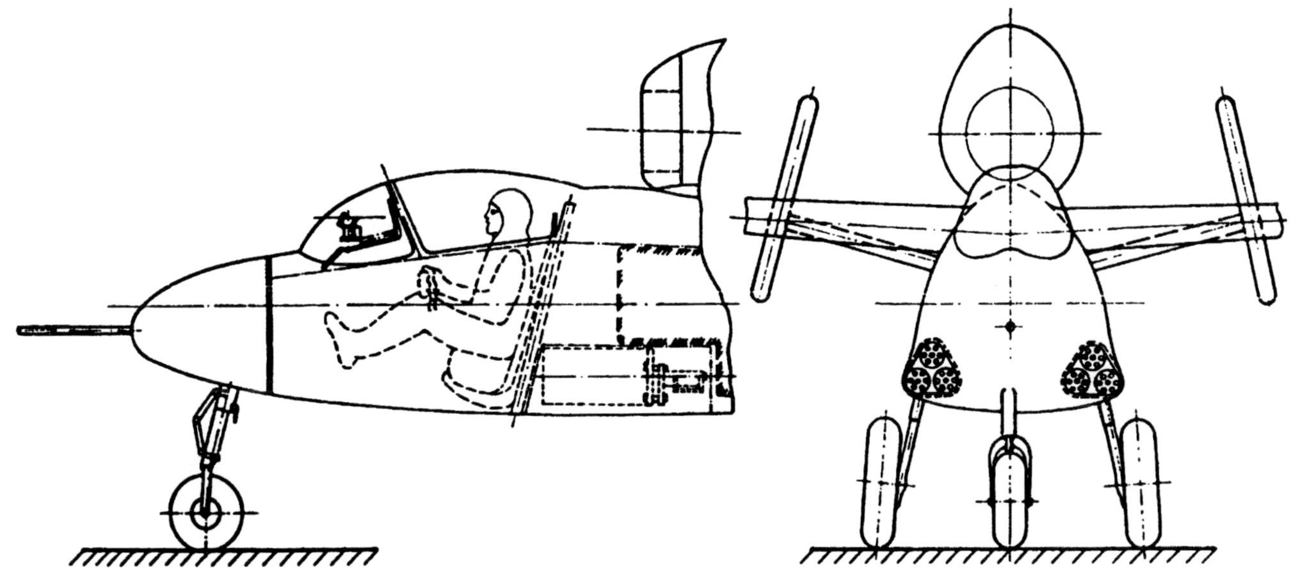

Schematic drawing of a Rohrblock battery installed in the He 162 in place of two MG 151/20 or MK 108. This armament system was developed by Rheinmetall-Borsig for use against heavy bombers.

Extract from the interim operating manual (8-162 A-2 and A-2) which was issued at the end of 1944.

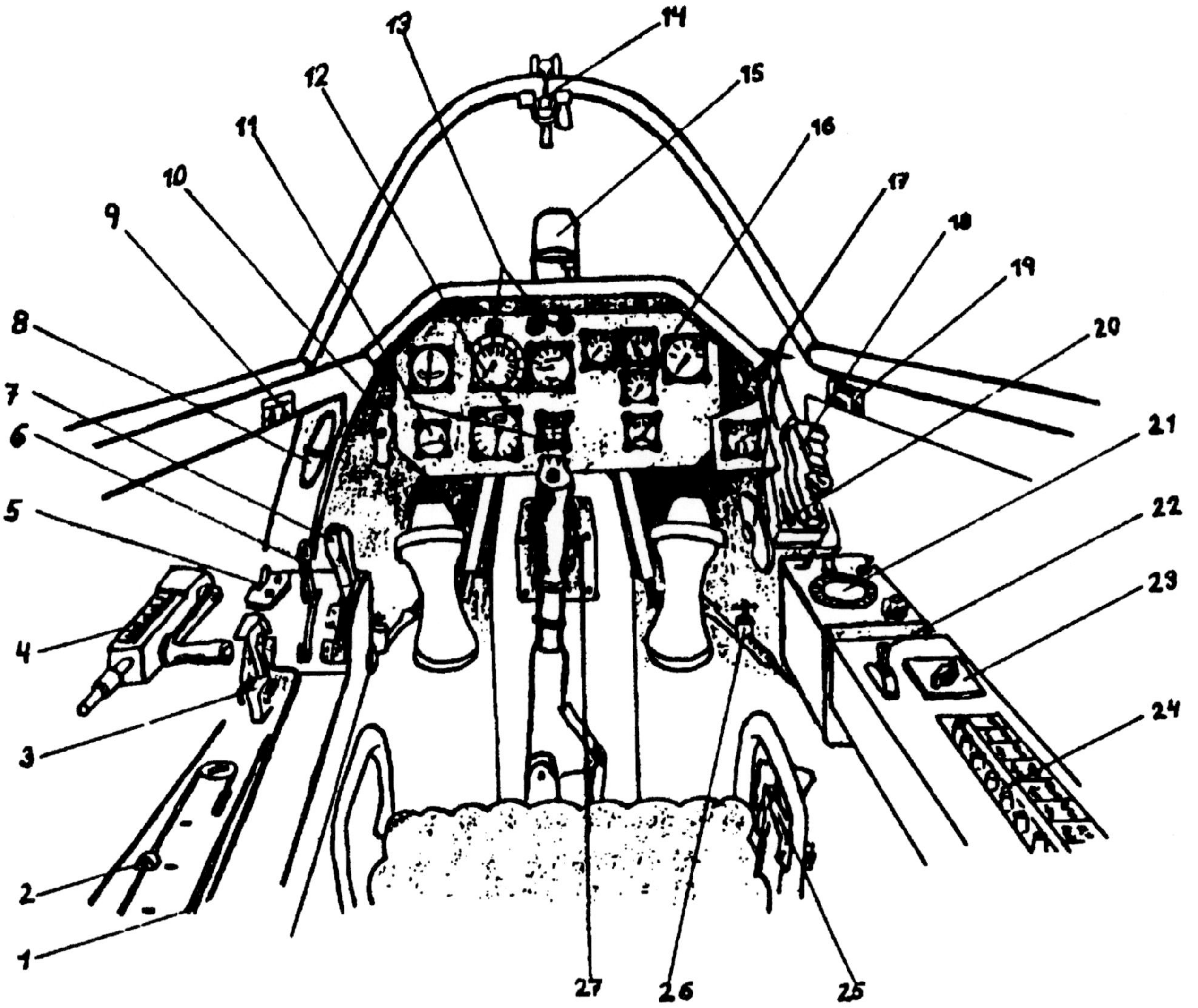

1 Takeoff assist rocket jettison
2 Rudder trim
3 Undercarriage lever
4 Elevator trim
5 Takeoff assist rocket ignition switch
6 Fuel cock lever
7 Throttle lever
8 Undercarriage lever
9 Undercarriage indicator left
10 Landing flap hand pump
11 Navigational equipment
12 Flight gauges
13 Airspeed indicator.
14 Canopy lock
15 Gunsight
16 Engine instruments
17 Oxygen system
18 Flare pistol
19 Undercarriage indicator right
20 FuG 25a panel
21 FuG 24 receiver
22 Starter switch
23 Variable nozzle switch
24 Circuit breakers
25 Ejection seat firing handle
26 Rudder pedal adjuster
27 Nosewheel well window

Circuit breakers:

Battery
Fuel pump
Weapons, gunsight
Pitot heat
Starter system and instruments
Artificial horizon, takeoff assist rockets
Variable nozzle
FuG 24
FuG 25a
Starter

The He 162 Cockpit

The restored and almost complete cockpit of the He 162 in the Musee de'l Air in Le Bourget, Paris.

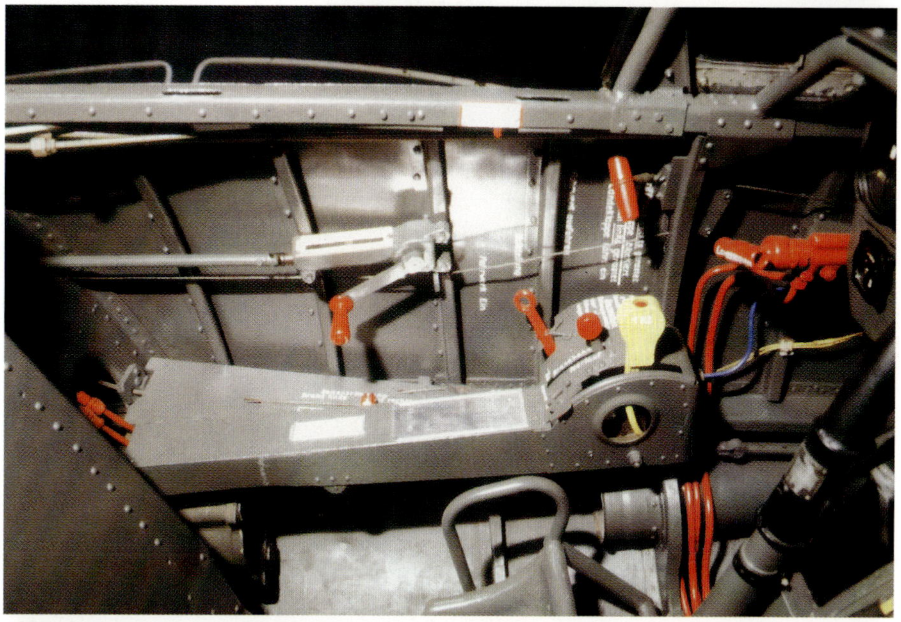

Left console with engine controls.

Right console. Weights have been installed in the cockpit to compensate for the cannon, which have been removed.

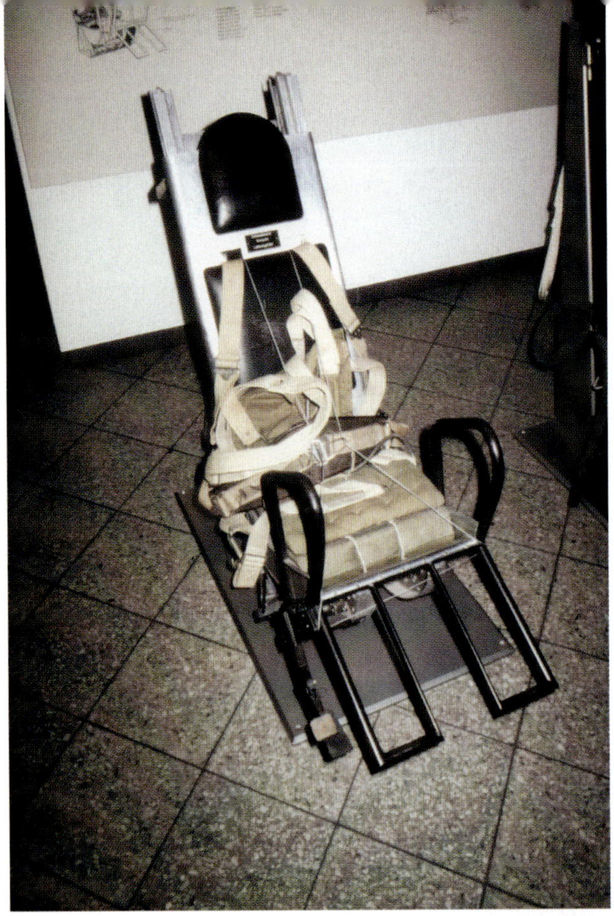

Above, Left: Heinkel He 162 ejection seat.

Above, Right: The aircraft at Le Bourget has weights placed in the cockpit to maintain the machine's normal center of gravity.

The instrument panel of the Heinkel He 162.

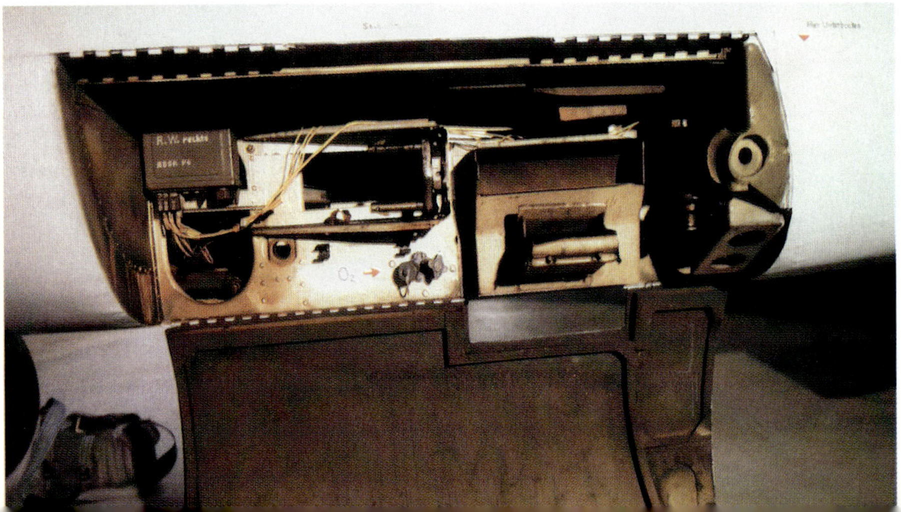

Weapons bay with the cannon removed.

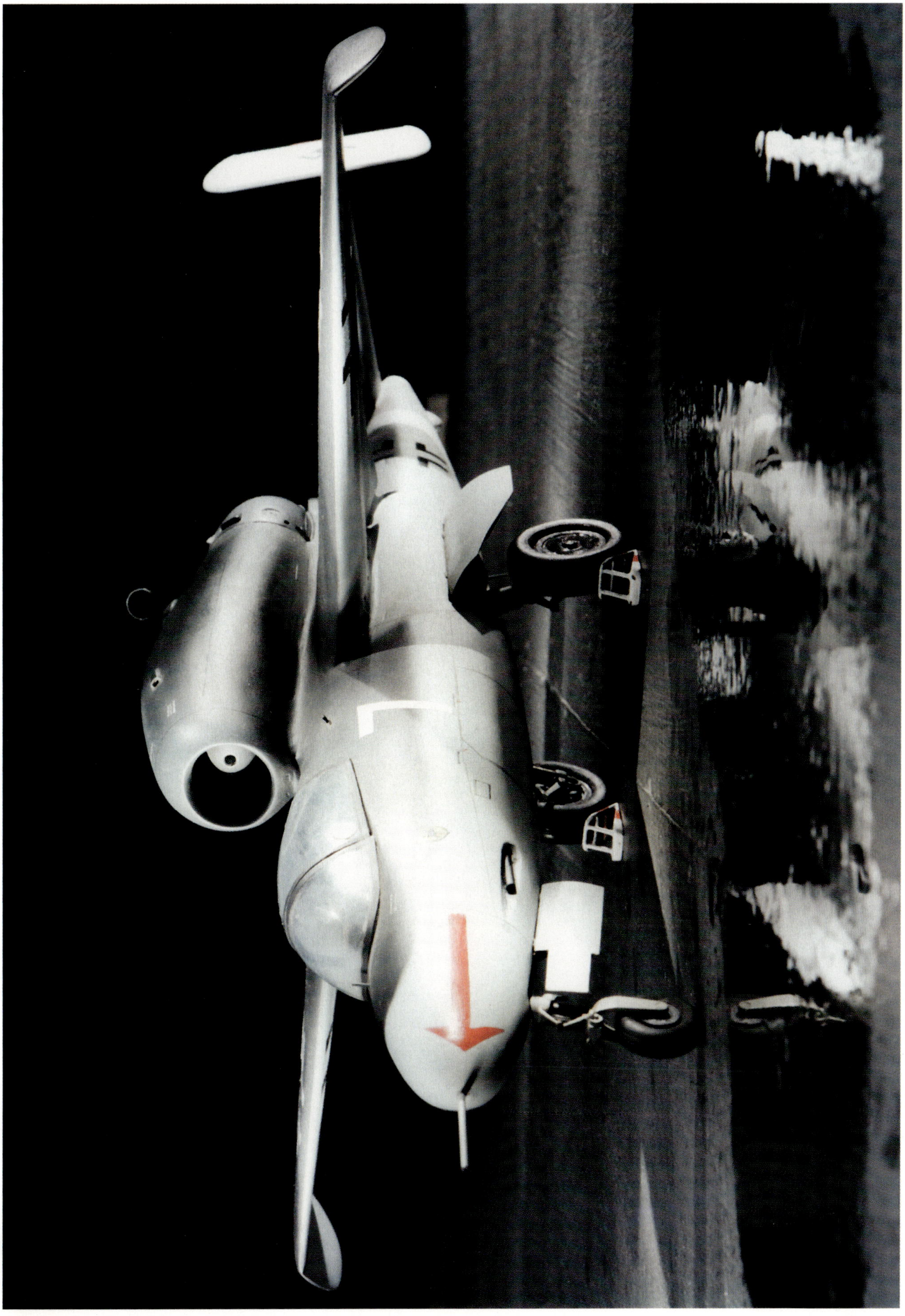

Realistic photo of Günter Sengfelder's model, taken by Robert Kroeschel.

Heinkel He 162 A-2
Scale 1 : 72

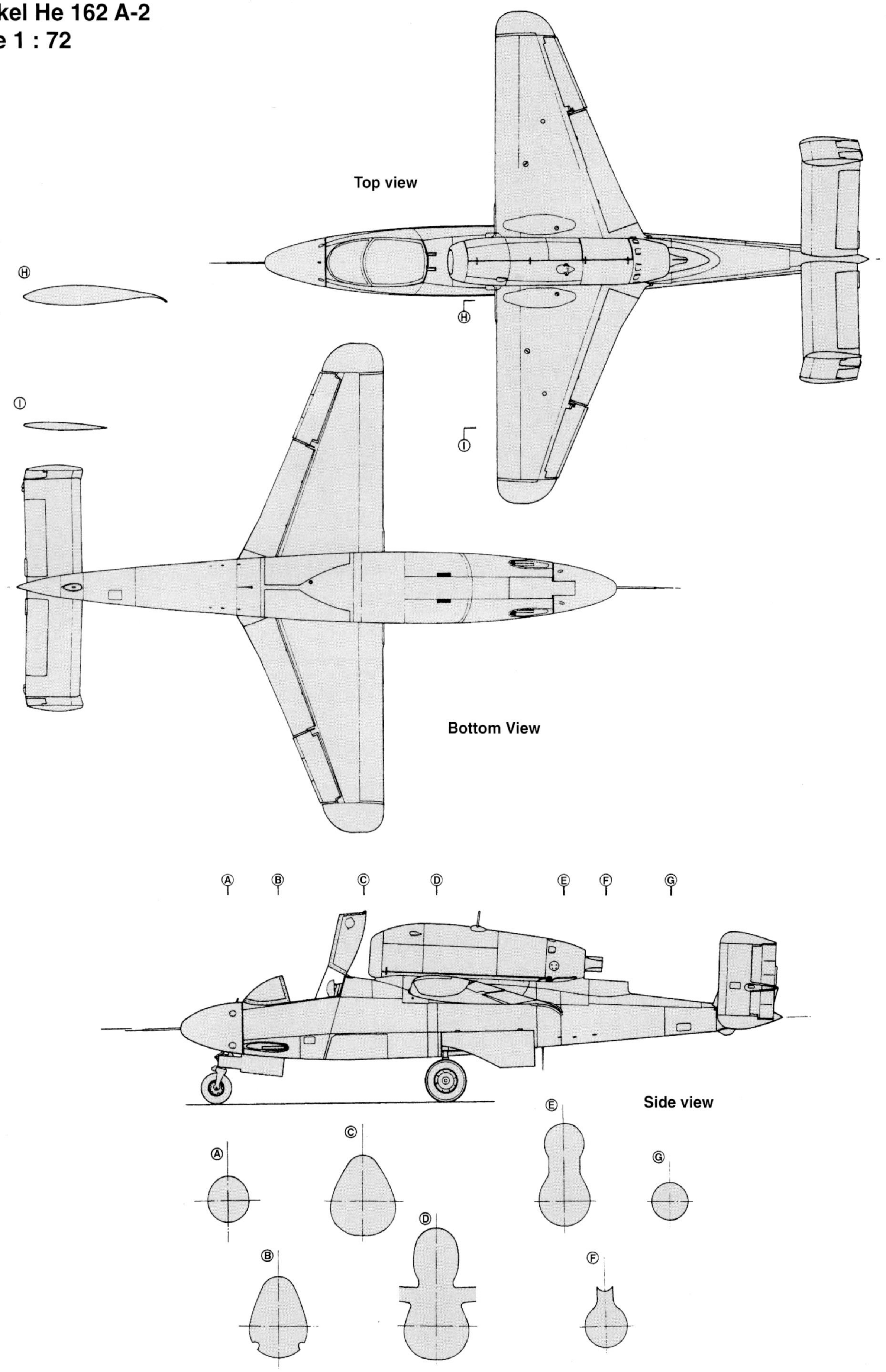